The Berry Cookbook

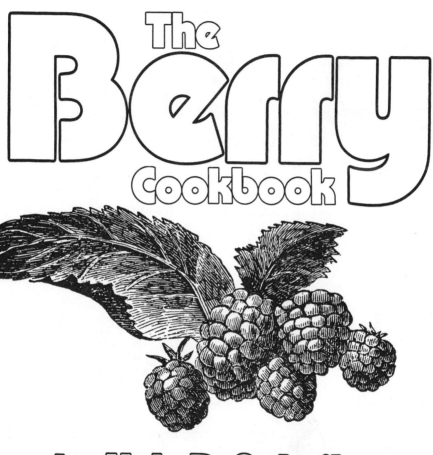

by Kyle D. Fulwiler

Pacific Search Press

To my great aunt Agnes Fulwiler

Pacific Search Press, 222 Dexter Avenue North,
    Seattle, Washington 98109
© 1981 by Kyle D. Fulwiler. All rights reserved
Printed in the United States of America

Designed by Judy Petry

Library of Congress Cataloging in Publication Data

Fulwiler, Kyle D.
    The berry cookbook.

    Includes index.
    1. Cookery (Berries) I. Title.
TX813.B4F84   641.6'47   81-1877
ISBN 0-914718-59-2   AACR2

Contents

Berries

When you think of berries, the first things that come to mind are jams, jellies, and pies. But if a blackberry thicket is threatening to take over your backyard, or your strawberry plants are so healthy the runners are really running, you need a wider scope of recipes to make use of the abundance nature has provided. And for those who don't have their own berry patches, there are many U-Pick farms, where for a bargain price you can pick enough berries to satiate your summer appetite for succulent fresh berries and still have enough left over to freeze or can.

This cookbook is designed to make it easy to use such a profusion of berries without getting bored. It provides recipes that range from refreshing beverages to elegant main dishes and light dessert soups, as well as the more usual fare, such as cobblers, crisps, muffins, and cakes. Almost all of these recipes work very well with either fresh or frozen berries, too (for the few recipes that require them, fresh berries are specified in the ingredients list), so you can enjoy your favorite berry desserts even in the middle of winter.

Preparing and Preserving Berries

Before you can or freeze berries or use them in recipes, be sure to wash them, remove the stems, and discard berries that have gone bad. Berries can be frozen up to one year. When packaging them for freezing, it is wise to measure them and record the amount on a label on the container so you need not remeasure them before using them in a recipe. Always leave a half inch of headspace in the container.

Although it may be more trouble than it's worth, canning may be your only option if a freezer is not available. Drain all canned berries well before using them in recipes.

If you want to make single-berry jams and jellies quickly and easily, use the specific directions on boxed pectin, which is available in most grocery stores. When making jams, jellies, or preserves, use jars that have been sterilized in boiling water for ten minutes or in a 225° oven for thirty minutes. Be sure to keep these jars hot until you are ready to use them. Use self-sealing lids that fit the jars and follow the manufacturer's directions for proper sealing.

Blackberries, Boysenberries, and Loganberries

Because blackberries, boysenberries, and loganberries are so closely related, they can be used interchangeably or in any combination in the recipes in this book. For blackberry pie, I prefer the wild mountain blackberry because the seeds are so small. My preferences for syrups and jellies are the larger evergreen and Himalaya blackberries. Keep in mind, too, that loganberries have large seeds.

To freeze blackberries, boysenberries, and loganberries, pack them in freezer containers and freeze. Thaw frozen berries in the freezer container before using.

To can these berries, prepare a syrup by slowly bringing three cups sugar and four cups water to a boil in a medium saucepan. Meanwhile, pack the berries in hot, sterilized self-sealing jars. Fill the jars with boiling syrup to a half inch from the top and seal. (This syrup is enough for about eight pints of berries.) Submerge the jars in a boiling water bath and process for fifteen minutes.

Blueberries

To freeze blueberries, pack them in freezer containers or heavy plastic bags and freeze. Shake them occasionally during the first two hours of freezing to keep them from sticking together. When using frozen berries in a recipe, remove them from the freezer when you start the recipe and use the berries when they are still partially frozen.

To can blueberries, blanch them for five minutes in boiling water and then partially cool them with cold running water. Pack them in hot, sterilized self-sealing jars, leaving a half inch of headspace, and seal. Place the jars in hot water, bring the water to a boil, then process the jars for fifteen minutes in the boiling water bath.

Cranberries

Cranberries are most commonly found during the holiday

season in your grocery store's produce department. These berries can be frozen up to one month in their original bags; if they will be frozen longer, wrap them in another heavy plastic bag to ensure freshness. You also can freeze them by measuring the amounts needed into freezer containers or heavy plastic bags. To use frozen berries, remove them from the freezer one hour before you need to use them in a recipe.

To can cranberries, make a syrup of four cups water and four cups sugar in a large saucepan, heating over medium heat until the sugar is dissolved. Bring the syrup to a boil, add cranberries, and boil for three minutes. (This syrup is enough for nine pints of berries.) Pack in hot, sterilized self-sealing jars, leaving a half inch of headspace. Process in a boiling water bath for ten minutes.

Gooseberries

To freeze gooseberries, remove the top and tail from each berry. Pack them in heavy plastic bags or freezer containers and freeze. Thaw frozen berries in the freezer container, then use as desired.

To can gooseberries, make a syrup by combining four cups water and three cups sugar in a large saucepan, heating over medium heat until sugar is dissolved. Bring the syrup to a boil, then add the berries and boil for thirty seconds. (This syrup is enough for about eight pints of berries.) Place in hot, sterilized self-sealing jars, leaving a half inch of headspace. Process in a boiling water bath for fifteen minutes.

Huckleberries

To freeze huckleberries, place them in heavy plastic bags and freeze. Shake the bag occasionally for the first two hours of freezing to keep berries from freezing in lumps. To use frozen huckleberries, remove them from the freezer when you start the recipe and use them partially thawed.

To can huckleberries, follow the directions given for gooseberries, but process them in a boiling water bath for twenty minutes.

Lingonberries

Fresh lingonberries are hard to find. Lingonberries are most commonly found in the gourmet sections of grocery stores as canned berries or as preserves. They have an interesting flavor that is reminiscent of that of cranberries; in fact, they are sometimes referred to as lowbush cranberries. If you are lucky enough to have a supply of lingonberries, you can freeze or can them in the same way as huckleberries.

Raspberries

To freeze raspberries, pack them in freezer containers and freeze. Before using them in recipes, let them thaw in the container.

To can raspberries, follow the instructions given for blackberries, boysenberries, and loganberries.

Strawberries

To freeze strawberries, hull all the berries and slice them or leave them whole in amounts to meet your anticipated needs. To freeze whole berries, spread them out on a cookie sheet and lay it flat in the freezer until the berries are frozen. Remove them from the freezer, pack them in freezer containers, and return them to the freezer. This method ensures that the berries hold a nice shape. To freeze sliced or puréed berries, put them directly into freezer containers and freeze. Thaw all strawberries in the freezer containers before using them in a recipe.

To can strawberries, use only whole, hulled berries. Using one cup sugar to each quart of berries, alternate layers in a shallow pan and let stand for one to two hours. Pour juice and berries into a large saucepan and heat until the mixture just comes to a boil; simmer for five minutes. Pack in hot, sterilized self-sealing jars, leaving a half inch of headspace, and seal. Process for fifteen minutes in a boiling water bath.

Appetizers, Beverages, and Garnishes

Summer Appetizers

Thinly sliced ham   1 4-ounce package
Large fresh strawberries   24, hulled
Sour cream   1 cup
Lemon peel   ½ teaspoon grated
Almonds   ¼ cup finely chopped and browned

Cut ham into 3 by ¾-inch strips. Wrap 1 strip of ham around base of each berry and secure with a toothpick. Combine sour cream and lemon peel in a small bowl. Dip the tip of each berry into cream mixture, then into almonds. Arrange coated berries on serving platter. Makes 24 appetizers.

Blackberry-Banana Shake-Up

Plain yogurt   1 cup
Banana   1, cut into chunks
Blackberries   ½ cup
Honey   1 tablespoon
Vanilla   1 teaspoon

Combine all ingredients in blender and blend for about 30 seconds. Serves 1.

Old-Fashioned Blackberry Cordial

Blackberries   2 quarts
Water   ½ cup
Sugar   1 cup
Brandy   2 cups

Combine berries and water in a large saucepan. Mash the berries while bringing the mixture to a boil over high heat; reduce heat and simmer until berries are soft. Remove from heat and pour into a jelly bag; let drip until all juice is extracted—you should have 4 cups. Pour juice and sugar into pan and simmer until mixture is slightly thickened and sugar is dissolved. Remove from heat and cool. Add brandy and bottle. Store in a cool place for 2 weeks before drinking. Makes 2 quarts.

Loganberry Refresher

Sugar ½ cup
Water ½ cup
Loganberries 7 cups
Orange juice 2 cups
Lemon juice ½ cup
Cold water 1½ cups

Combine sugar and water in a small saucepan and cook over medium high heat until sugar is dissolved, stirring constantly; remove from heat and chill. Mash berries, then press through a strainer to remove seeds. Combine berries with orange juice and lemon juice, and add sugar syrup. Chill until very cold. Before serving, add cold water. Serves 8.

Cranberry Sparkle

Small ice block 1 or
   Ice cubes about 12
Cranberry juice 1 quart, chilled
Orange juice 1 cup, chilled
Ginger ale 1 quart, chilled
Orange ½, sliced
Lime 1, sliced
Maraschino cherries 1 4-ounce jar

Place ice in a punch bowl. Add cranberry juice and orange juice. Slowly pour in ginger ale. Float orange and lime slices and cherries on top of punch. Serves 12.

Holiday Punch

Grapefruit juice   1 quart, chilled
Lemon juice   ½ cup
Cranberry juice   1 cup, chilled
Vodka   1 fifth, chilled
Limes   2, quartered
Orange   1, thinly sliced
Soda water   1 quart, chilled

Combine fruit juices and vodka in a very large punch bowl. Squeeze juice from lime quarters into punch bowl, then add quarters to punch. Add orange slices, and pour in soda water. Serves 20.

Bright Brunch Punch

Raspberries   3½ cups
Sugar   to taste
Orange liqueur   ¾ cup
Orange flower water   6 drops
Champagne   2 bottles, chilled

Purée berries in a food processor or smash with a large spoon; press through a strainer to remove seeds. Add sugar, liqueur, and flower water. Stir well. Chill for at least 2 hours and add champagne. Serves 8.
Note This drink looks very attractive served in tall long-stemmed wine glasses.

Johnstons' Favorite Punch

Frozen orange juice concentrate   2 12-ounce cans, thawed
Frozen lemonade concentrate   1 12-ounce can, thawed
Water   4½ quarts
Raspberries   1 quart
Bananas   2, peeled
Ice cubes   3 cups

Combine orange juice, lemonade, and water in a very large punch bowl (about 2-gallon capacity) and mix well. Purée berries and bananas in a blender or food processor, adding a little water if necessary to mix. Add to mixture in punch bowl and stir until com-

bined; add ice. Serves 30.
Variation Add 1 quart of pineapple juice or 7-Up.

Raspberry Champagne Bowl

Port wine   1 fifth
Raspberries   1 quart
Sugar   to taste
Lime   1, sliced
Champagne   2 fifths, chilled

Combine wine, berries, sugar, and lime slices in a large saucepan. Bring to a boil over high heat, then reduce heat and simmer for 8 minutes. Remove from heat and strain into a large bowl. Cover and chill. When cold, pour into a punch bowl and add champagne. Serve immediately. Serves 20.

Raspberry Syrup

Raspberries   5 pounds
Sugar   8 cups

Crush berries, press through a fine sieve, and pour into a large saucepan. Add sugar and slowly bring to a boil, stirring constantly. Boil for 2 minutes and remove from heat. Let stand 5 minutes, then return to heat, slowly bring to a boil, and boil for 3 minutes more. Skim off any foam that forms on top of syrup. Pour syrup into a hot, sterilized self-sealing quart jar. This syrup may be used immediately after cooling, or it can be stored in a cool, dark place if the lid is sealed securely. Makes 1 quart.
Note Raspberry syrup can be mixed to taste with champagne, soda water, or plain water for a refreshing drink. It also is great as a pancake topping.

Raspberry Vodka

Raspberries   3 pounds
Lemon peel   of 1 lemon, grated
Sugar   2 cups
Vodka   1 fifth

Place berries, lemon peel, and sugar in a large kettle. Bring to a boil, then reduce heat and simmer 3 minutes. Remove from heat and cool to room temperature. Push mixture through a wire strainer to remove seeds and lemon peel. Add vodka to strained mixture and pour into bottles. Store in a cool, dark place for 2 months before drinking. Makes 1½ quarts.
Note Mix with tonic water and a slice of lime for a refreshing summer drink.

Breakfast in a Blender

Any flavor of yogurt   ¾ cup
Ice cubes   3
Honey   1 tablespoon
Egg   1
Strawberries   1 cup, hulled

Combine all ingredients in a blender and blend until smooth. Serves 1.

Strawberries in Champagne

Strawberries   2 cups, hulled
Champagne   1 bottle, chilled

Place berries in 4 6-ounce wine glasses. Pour champagne over berries and serve. Serves 4.
Note This is a great way to start off a very special Sunday brunch.

Strawberry Daiquiri

Rum 1½ ounces
Lime juice 1 tablespoon
Sugar 2 teaspoons
Strawberries ¼ cup hulled and sliced
Ice cubes 6

Combine rum, lime juice, sugar, and berries in a blender; blend on high speed until berries are pulverized. Add ice cubes and blend until mixture is slushy. Serves 1.

Strawberry Margarita

Lime 1
Salt
Tequila 1 cup
Frozen limeade concentrate 1 6-ounce can, thawed
Triple Sec ¼ cup
Strawberries 5, hulled
Lemon juice 1 tablespoon
Crushed ice 2 cups

Cut lime in half and rub rims of 6 5-ounce glasses with cut side. Dip each rim in salt. Combine remaining ingredients in blender and mix on high speed. Pour into glasses. Serves 6.

Strawberry Piña Colada

White rum 1½ ounces
Cream of coconut 1½ ounces
Pineapple juice 1 ounce
Strawberries 4, hulled
Sugar to taste
Crushed ice ½ cup

Combine all ingredients in blender and blend on high speed for 1 minute. Pour into a 12-ounce glass. Serves 1.

Strawberry Rosé Punch

**Frozen strawberries*   2 10-ounce packages, thawed
Frozen lemonade concentrate   2 6-ounce cans, thawed
Rosé wine   ½ gallon
7-Up   7 12-ounce bottles

Blend berries and lemonade in a blender or food processor until smooth. Add wine and 7-Up. Pour into punch bowl, add ice cubes or ice mold, and serve. Serves 25.
* This can be made with fresh strawberries by blending 2 cups sliced berries and ½ cup sugar with lemonade.

Cranberry-Lemon Garnish

Lemons   4
Cranberry Ice (see Index)   1 quart

Cut lemons lengthwise and scoop out pulp with a grapefruit knife. (Pulp can be pressed through a strainer to reserve juice for another purpose.) Make a slight cut on the bottom of each shell so it will stand upright. Fill each with ½ cup of cranberry ice, cover, and freeze until firm. Makes 8 garnishes.
Note This is an excellent garnish on a bed of parsley around a roasted turkey.

Preserves and Sauces

Blueberry-Apple Butter

Blueberries   1 quart
Apples   4, peeled, cored, and chopped
Sugar   4 cups
Ground allspice   1 teaspoon
Mace   1 teaspoon
Cinnamon   ½ teaspoon
Ground ginger   ¼ teaspoon

Combine all ingredients in a large kettle and stir until well mixed. Bring to a boil over medium high heat, stirring constantly; lower heat and simmer for 45 minutes, or until mixture is very thick (about the consistency of thick applesauce). Pour into hot, sterilized self-sealing jars. Makes 3 pints.

Blueberry Jam

Blueberries   1 quart
Sugar
Apple   1, peeled, cored, and chopped

Put half the berries in a large saucepan and crush slightly; add remaining berries. Cook over medium heat until soft. Remove from heat and measure; add 1 cup sugar for each cup berry pulp. Return to pan and add apple. Stir until sugar dissolves. Bring to a boil over medium heat; reduce heat and cook until thick. To test, put a spoonful on a plate; if it doesn't run, it is thick enough. Pour into hot, sterilized self-sealing jars. Makes 1½ pints.

Blueberry-Cherry Conserve

Sour cherries*   3 cups, pitted
Water   ¼ cup
Blueberries   3 cups
Sugar   4½ cups

In a large kettle, cook cherries with water over medium heat just until skins are tender. Add berries and sugar and stir gently until sugar is dissolved; cook over medium high heat about 30 minutes, or until thick and clear. Pour into hot, sterilized self-sealing jars. Makes 2½ pints.

* If fresh cherries are not available, use canned pitted tart cherries packed in water.

Cranberry Catsup

Cranberries   2½ pounds
Vinegar   5 cups
Sugar   2⅔ cups
Cinnamon   1 tablespoon
Ground cloves   1 teaspoon

Put berries in a large saucepan and cover with vinegar. Bring to a boil; reduce heat and simmer about 10 minutes, or until berries pop. Push mixture through a metal strainer to remove berry skins. Return mixture to pan and add remaining ingredients; cook over medium heat until thick. Pour into hot, sterilized self-sealing jars. Makes 3 pints.

Note Serve this instead of cranberry sauce with roasted poultry.

Cranberry Chutney

Cranberries   1 1-pound package
Dates   ¾ cup pitted and quartered
Water   ½ cup
Sugar   ½ cup
Vinegar   ¼ cup
Ground ginger   ¼ teaspoon
Nutmeg   ¼ teaspoon
Ground allspice   ⅛ teaspoon
Ground cloves   ⅛ teaspoon
Salt   ⅛ teaspoon
Golden raisins   ¾ cup

Combine all ingredients except raisins in a large kettle and bring to a boil over high heat; reduce heat and simmer 5 minutes, stirring constantly. Add raisins and continue simmering and stirring 5 minutes more. Pour into small, hot, sterilized self-sealing jars. Makes 1½ pints.

Cranberry-Orange Relish

Orange   1
Cranberries   2 cups
Sugar   1 cup

Quarter orange and put through a food chopper or food processor until finely chopped. Combine all ingredients in a large bowl and stir until blended. Refrigerate overnight before using. Makes 2 cups.

Cranberry-Peach Relish

Water   ½ cup
Vinegar   ½ cup
Ground cloves   1 teaspoon
Cinnamon stick   1 2-inch piece
Ground ginger   ⅛ teaspoon
Sugar   1 cup
Canned peach halves   6
Cranberries   2 cups

Combine first 6 ingredients in a large saucepan and heat to boiling; remove cinnamon stick. Add peaches and berries and simmer about 10 minutes or until berries pop. Remove from heat, cover, and chill until very cold. Makes about 4 cups.

Gooseberry Catsup

Gooseberries   3 quarts
Hot Water
Sugar   5 cups
Nutmeg   ½ teaspoon
Ground cloves   1½ teaspoons
Cinnamon   1 teaspoon
Vinegar   2 cups

Scald berries with hot water; drain. Push through a fine strainer to remove seeds and skins; put in a large kettle. Stir in sugar, nutmeg, cloves, and cinnamon. Bring to a boil over medium high heat, stirring constantly; boil 15 minutes. Remove from heat and add vinegar.

Pour into hot, sterilized self-sealing jars. Store in a cool, dark place. Makes 3 pints.

Note Serve this with stews and roasted meats.

Gooseberry Jam

Gooseberries   2 quarts
Water   1 cup
Sugar

In a medium saucepan, cook berries and water over medium heat until berries are soft. Measure cooked berries and add ¾ cup sugar for each cup of berries. Boil gently until very thick. To test, put a small amount on a plate; if it doesn't run, it is thick enough. Pour into hot, sterilized self-sealing jars. Makes 3 pints.

Spiced Gooseberry Relish

Gooseberries   3 quarts
Sugar   4 cups
Nutmeg   ½ teaspoon
Ground cloves   1 tablespoon
Cinnamon   1 teaspoon
Apple cider vinegar   1 cup

Combine all ingredients in a large saucepan and stir well. Bring mixture to a boil; cover and simmer for 45 minutes, stirring occasionally. Pour into hot, sterilized self-sealing jars. Makes 4 pints.

Huckleberry Jam

Huckleberries   2 quarts
Lemon juice   2 tablespoons
Fruit pectin   1 1¾-ounce box
Sugar   7 cups

Crush berries completely and add lemon juice. Pour into a large kettle and add fruit pectin. Cook according to directions enclosed in pectin box for jam. Pour into hot, sterilized self-sealing jelly jars and allow to cool. Store in a cool place. Makes 4¼ pints.

Huckleberry and Strawberry Jam

Huckleberries   1 pound
Strawberries   1 quart, hulled
Lemon juice   2 tablespoons
Fruit pectin   1 1¾-ounce box
Sugar   5½ cups

Crush huckleberries and strawberries together; add lemon juice and pectin and pour into a large kettle. Cook according to directions in pectin box for jam. Pour into hot, sterilized self-sealing jelly jars and allow to cool. Store in a cool place. Makes 3½ pints.

Lingonberry Relish

Lingonberries   2 cups
Apple   1, peeled and cored
Orange   1
Lemon peel   2 teaspoons grated
Sugar   1 cup

Grind berries and apple in a food grinder or processor and put in a medium bowl. Peel orange and then remove white pith. Chop orange peel and pulp; add with lemon peel to berry mixture. Stir in sugar. Cover and chill in refrigerator overnight. Makes 1 quart.
Note Serve this relish as you would cranberry sauce.

Raspberry-Blueberry Jelly

Blueberries   1 quart
Raspberries   1 cup
Lemon juice   ¼ cup
Water   1½ cups
Sugar   5 cups
Fruit pectin   1 1¾-ounce package

Combine berries, lemon juice, and water in a large kettle and bring to a boil; reduce heat and simmer 5 minutes. Place mixture in jelly bag and squeeze out juice; there should be 4 cups. Follow directions enclosed in pectin box for cooking jelly. Pour into hot, sterilized self-sealing jars. Makes 3½ pints.

Magic Strawberry Preserves

Hot water
Strawberries   2 quarts, hulled
Sugar   4 cups
Sugar   2 cups

Pour hot water over berries and let stand 2 minutes; drain. Put berries in a large saucepan and add 4 cups sugar; bring to a boil over medium high heat, stirring constantly. Boil 2 minutes. Remove from heat and add remaining sugar. Return to heat and boil 5 minutes more. Pour mixture into a shallow pan and let stand overnight, stirring occasionally. Pour into jelly jars, and seal with paraffin or freeze. Makes 3 pints.

Strawberry Conserve

Strawberries   5 quarts, hulled
Orange peel   of 3 oranges, grated
Orange juice   of 3 oranges
Lemon peel   of 1 lemon, grated
Sugar   ¾ of weight of berries
Almonds   1½ cups, blanched and quartered

Weigh berries to determine the amount of sugar needed. Combine berries, orange peel, orange juice, and lemon peel in a large kettle. Pour sugar over this mixture and mix well. Let stand overnight. Cook over medium high heat for 35 minutes, stirring occasionally to prevent sticking. Add almonds and continue cooking 10 minutes more. Pour into hot, sterilized self-sealing jars. Makes 7½ pints.

Apple-Blueberry Sauce

Blueberries   2 cups
Water   ½ cup
Brown sugar   ¼ cup
Sugar   ¼ cup
Apple   1, peeled, cored, and roughly chopped
Orange   1, peeled and roughly chopped
Cinnamon   ¼ teaspoon
Pecans   ½ cup chopped

In a medium saucepan, combine berries and water and bring to a boil. Add next 5 ingredients and cook for about 10 minutes over medium high heat. Remove from heat and add pecans. Pour into a bowl, cover, and chill well. Makes 3 cups.
Note This is good served with a pork roast.

Low-Cal Blueberry Sauce

Blueberries   2 cups
Water   ⅔ cup
Cornstarch   1½ tablespoons
Liquid artificial sweetener   2 teaspoons
Salt   a dash
Lemon peel   1 teaspoon
Lemon juice   1 tablespoon

In a medium saucepan, bring first 6 ingredients to a boil over medium heat and then simmer 5 minutes, stirring constantly. Remove from heat and stir in lemon juice. Store covered in refrigerator. Makes 1½ cups.
Note This sauce may be served hot or cold and is delicious served over ice cream.

Cranberry Sauce

Cranberries   1 1-pound package
Water   1 cup
Sugar   2 cups

In a large saucepan, cook berries with water over high heat until berries pop open. Reduce heat to medium and add sugar. Cook, stir-

ring constantly, about 15 minutes, or until sauce is very thick. Makes
4 cups.

Cran-Orange Sauce

Sugar 2 cups
Water 1 cup
Cinnamon stick 1 2-inch piece
Whole cloves 6
Whole allspice 4
Cranberries 1 quart
Orange peel 2 teaspoons grated or
    Dried orange peel 2 teaspoons

Combine first 5 ingredients in a large saucepan. Bring to a boil over
medium heat, stirring constantly; reduce heat and simmer for
5 minutes. Remove and discard cloves and allspice. Add berries;
simmer uncovered for about 15 minutes, or until berries pop.
Remove from heat and discard cinnamon stick. Stir in orange peel.
Cool, then cover and chill. Makes 4 cups.
Note This freezes very well.

Cranberry-Pear Sauce

Cranberries 1 pound
Pears 2½, peeled, cored, and diced
Golden raisins 1 cup
Sugar 1 cup
Fresh orange juice ½ cup
Orange peel 1 tablespoon grated
Cinnamon 1 teaspoon
Nutmeg a dash
Orange liqueur ¼ cup

Combine all ingredients except liqueur in a large saucepan and
bring to a boil. Reduce heat and simmer, stirring frequently, about
25 minutes, or until mixture thickens. Remove from heat and stir in
liqueur. Cover and refrigerate for at least 6 hours. Serve this sauce
slightly chilled. Makes 5 cups.
Note This can be preserved in self-sealing jars; process in a hot
water bath for 5 minutes.

Gooseberry Sauce

Gooseberries   2 cups
Water   ¼ cup
Sugar   ¾ cup plus 2 tablespoons
Cornstarch   2 tablespoons
Water   3 tablespoons

In a medium saucepan, cook berries, water, and sugar over medium heat until berries are soft. Mix together cornstarch and water and add to berries; continue cooking until mixture thickens. Makes 2 cups.

Lingonberry Sauce

Apples   2½ cups peeled, cored, and chopped
Water   ½ cup
Sugar   1½ cups
Lingonberries   1 quart

In a medium saucepan, cook apples and water over medium heat until apples are soft. Add sugar and berries and continue cooking about 15 minutes, or until mixture is thick. Makes 1 quart.
Note Serve with roasted meats or Swedish pancakes.

Lingonberry Spice Sauce

Lingonberry preserves   1 14¾-ounce jar
White wine   ¼ cup
Orange juice   ¼ cup
Lemon juice   1 tablespoon
Orange peel   1 tablespoon grated or
    Dried orange peel   1 tablespoon
Cinnamon   1 teaspoon
Ground allspice   ½ teaspoon

Combine all ingredients in a medium saucepan and bring to a boil over high heat; reduce heat and simmer for 15 minutes. Cover and chill. Makes 2 cups.
Note Serve with roasted meats.

Raspberry Wine Sauce

Raspberries 2 cups
Cranberry juice 1½ cups
Sugar ¾ cup
Cornstarch 3 tablespoons
Red wine ½ cup
Cinnamon ¼ teaspoon

Combine berries and cranberry juice in a medium saucepan and bring to a boil. Push mixture through a strainer and discard pulp; return berries to saucepan. Combine sugar and cornstarch and add to berry mixture; cook over medium high heat until thick, stirring constantly. Remove from heat and add wine and cinnamon. Cover and chill. Makes 2¾ cups.
Note This is delicious served with a variety of sherbet balls.

Fruit Salad Sauce

Strawberries 1½ cups, hulled
Frozen raspberries 1 10-ounce package, thawed and drained
Frozen orange juice concentrate 1 tablespoon thawed
Strawberries ½ cup hulled and sliced

Combine all ingredients except ½ cup strawberries in a food processor or blender and purée. Pour into a small bowl. Stir in sliced strawberries and chill well before serving. Makes 2 cups.
Note This is very good served over a variety of sliced summer fruits.

Breakfasts, Brunches, and Breads

Swiss Cold Cereal

Quick-cooking oats   2 tablespoons
Cream   2 tablespoons
Sugar   1 tablespoon
Apple   1, cored and grated
Lemon juice   1 teaspoon
Raspberries   1 cup
Whipping cream   2 tablespoons whipped
Nuts   1 tablespoon chopped

Combine oats, cream, and sugar and chill for 2 hours. Stir in apple and lemon juice. Fold in berries, whipped cream, and nuts. Serves 1.
Note This may be served as a dessert, snack, or breakfast treat.

Blueberry Pancakes

Flour   1 cup
Soda   ½ teaspoon
Baking powder   1 teaspoon
Salt   ½ teaspoon
Whole wheat flour   ¼ cup
Egg   1
Milk   1¼ cups
Oil   2 tablespoons
Blueberries   1 cup
Oil   for frying
Butter   topping
Syrup   topping

Sift flour, soda, baking powder, and salt; add whole wheat flour and set aside. Mix egg and milk together in a medium bowl and add oil. Slowly stir in flour mixture. Let batter stand at room temperature for at least 20 minutes. Meanwhile, heat griddle or large frying pan with a thin film of oil. Drop large tablespoons of batter on griddle and add a few blueberries to top of pancakes. When set, flip over and cook other side. Serve with butter and syrup. Serves 4.
Note You can use any prepared mix to make blueberry pancakes; add blueberries and cook pancakes in same manner described above.

Swedish Pancakes with Lingonberries

Milk 2 cups
Flour ½ cup
Salt ½ teaspoon
Sugar 2 tablespoons
Eggs 4
Lingonberries 1 14½-ounce jar, heated
Butter topping

Combine milk, flour, salt, and sugar and beat until smooth. Add eggs and beat until well combined. Let mixture stand at room temperature for 30 minutes. To cook, pour a large tablespoonful onto an electric grill preheated to 400°. Cook until set and turn. These may also be cooked in an 8-inch frying pan in the same manner as crepes. Keep warm on a hot plate until serving time. Serve with lingonberries and butter. Serves 4.

Strawberry Waffles

Butter 6 tablespoons
Powdered sugar 1 cup
Vanilla ¼ teaspoon
Waffles 4 servings
Strawberries 1 cup hulled and sliced

In a small bowl, beat butter until smooth. Gradually add powdered sugar and continue beating until light and fluffy. Beat in vanilla. Spread creamed mixture over waffles and top with berries. Serves 4.

Blackberry Syrup

Blackberries*   2 quarts
Sugar   4 cups

Crush berries in a large kettle and heat slowly until fruit is very soft. Scald a jelly bag with hot water and wring out. Pour fruit into jelly bag and let drip for about 12 hours, or until all juice has strained through bag. Put juice and sugar in a large kettle and slowly bring to a full boil, stirring constantly; boil for 5 minutes. Pour into a sterilized quart-size long-neck bottle. Keep refrigerated; use on waffles and pancakes. Makes about 1 quart.

*The large evergreen blackberries are best for this recipe.

Low-Cal and Creamy Blueberry Toasts

English muffins   2, split
Ricotta cheese   8 tablespoons
Blueberries   2 cups
Artificial sweetener   to taste
Cinnamon   to taste
Nutmeg   to taste

Place muffin halves on a cookie sheet, split sides up. Spread 2 tablespoons cheese on each, sprinkle with berries, and sweeten to taste with sweetener, cinnamon, and nutmeg. Bake at 350° for 15 minutes. Serves 4.

Blueberry Blintzes

Cottage cheese   3 cups
Egg yolks   2, beaten
Butter   1 tablespoon, softened
Sugar   1 tablespoon
Cinnamon   ¼ teaspoon
Dessert Crepes (see Index)   14 to 16, cooked on only one side
Sugar   ½ cup
Arrowroot   2 teaspoons
Pineapple juice   ½ cup
Water   ½ cup
Blueberries   1½ cups
Lemon extract   ½ teaspoon
Butter   2 tablespoons

Push cottage cheese through a strainer or blend in a food processor until smooth. In a large bowl, combine cheese with egg yolks, butter, sugar, and cinnamon and blend well. Place one tablespoon of mixture on cooked side of each crepe, then fold bottom up, sides in, and top down; set crepes aside. Combine sugar, arrowroot, pineapple juice, water, and berries in a saucepan and cook over medium heat until thick. Remove from heat and add lemon extract. Fry filled crepes in butter over medium heat until golden brown. Serve topped with blueberry sauce. Makes 14 to 16 blintzes.

Huckleberry Fritters

Flour   1¾ cups
Baking powder   1 tablespoon
Salt   ½ teaspoon
Sugar   2 tablespoons
Eggs   2, beaten
Milk   1 cup
Margarine   1 tablespoon, melted
Huckleberries   1 cup
Oil   for frying

Sift flour, baking powder, salt, and sugar into a large bowl. Combine eggs, milk, and margarine and add to flour mixture; stir until well combined. Fold in berries. In a heavy saucepan, heat oil to 365°. Drop batter by teaspoonfuls into hot oil and fry until golden brown. Serve warm. Makes about 3 dozen fritters.

Blackberry Coffee Cake

Flour   2 cups
Sugar   1 cup
Baking powder   2 teaspoons
Salt   1 teaspoon
Cinnamon   ½ teaspoon
Margarine   ½ cup
Eggs   2
Milk   1 cup
Vanilla   1 teaspoon
Blackberries   3½ cups
Brown sugar   ⅓ cup
Flour   ¼ cup
Butter   2 tablespoons
Pecans   ½ cup chopped

Sift together first 5 ingredients and put in a large bowl; cut in margarine until mixture looks like crumbs. In another bowl, mix together eggs, milk, and vanilla. Pour over flour mixture and stir until just moistened. Spread in a greased 8 by 12-inch pan; distribute berries on top. In a small bowl, combine brown sugar, flour, and butter; mix with a fork until crumbly. Add pecans and sprinkle over top of cake. Bake at 350° for 45 minutes.

Huckleberry Coffee Cake

Margarine   ⅔ cup
Sugar   1¼ cups
Eggs   2
Orange peel   1½ teaspoons
Flour   2 cups
Baking powder   2 teaspoons
Salt   1 teaspoon
Milk   ¾ cup
Orange flavoring   1 teaspoon
Huckleberries   2 cups
Sugar   ¼ cup

Beat together margarine and sugar until well creamed. Add eggs 1 at a time, beating well after each addition. Stir in orange peel. Sift together flour, baking powder, and salt. Alternately mix sifted ingredients and milk into creamed mixture. Stir in orange flavoring and berries. Pour into a greased and floured 9 by 13-inch pan; sprinkle sugar over batter. Bake at 350° for 40 minutes.

Blueberry-Carrot Bread

Eggs 3
Oil 1 cup
Vanilla 2 teaspoons
Sugar 3 cups
Flour 3 cups
Cinnamon 1 teaspoon
Soda 1 teaspoon
Carrots 2 cups grated
Blueberries 1 cup

Beat together eggs and oil until well beaten. Add vanilla and slowly mix in sugar. Sift together flour, cinnamon, and soda; stir into egg mixture alternately with carrots and berries. Pour into 2 greased 9 by 5-inch loaf pans. Bake at 375° for 1 hour. Makes 2 loaves.

Blueberry Molasses Bread

Flour 1 cup
Whole wheat flour 2 cups
Sugar ½ cup
Salt 1 teaspoon
Molasses ½ cup
Soda 1 teaspoon
Milk 1½ cups
Blueberries 1½ cups

Combine flour, whole wheat flour, sugar, and salt in a large bowl. Pour in molasses. Combine soda and milk, stirring until soda is dissolved; slowly pour into flour mixture and stir until well mixed. Carefully fold in berries. Pour mixture into a greased and floured 9 by 5-inch loaf pan. Bake at 325° for 1½ hours. Makes 1 loaf.

Blueberry Nut Bread

Unbleached white flour   1½ cups
Sugar   1 cup
Baking powder   1½ teaspoons
Salt   1 teaspoon
Baking soda   ½ teaspoon
Whole wheat flour   ½ cup
Margarine   ¼ cup
Egg   1, well beaten
Orange juice   ¾ cup
Orange peel   1 teaspoon grated or
    Dried orange peel   1 teaspoon
Blueberries   1 cup
Walnuts   ½ cup chopped

Sift together first 5 ingredients and put in a large bowl; add whole wheat flour. Cut margarine into flour mixture. In a small bowl, combine egg, orange juice, and orange peel. Add to dry ingredients and mix until just moistened. Fold in berries and walnuts. Pour into a greased and floured 9 by 5-inch loaf pan and bake at 350° for 1 hour. Cool before removing from pan; wrap in foil and store overnight before cutting. Makes 1 loaf.

Blueberry-Pineapple Bread

Pineapple chunks packed in pineapple juice   1 15¼-ounce can,
    juice drained and reserved
Pineapple juice plus water   1 cup
Oil   2 tablespoons
Vanilla   1 teaspoon
Egg   1
Flour   2 cups
Sugar   1 cup
Baking powder   1 teaspoon
Salt   ½ teaspoon
Baking soda   ½ teaspoon
Almonds   ½ cup chopped
Blueberries   1 cup

Drain pineapple and measure juice; add water to juice to make 1 cup and set aside. In a large bowl, beat oil, vanilla, and egg until thoroughly mixed. Sift together next 5 ingredients; beat into egg mixture alternately with pineapple juice. Fold in pineapple chunks, almonds, and berries. Pour into a greased and floured 9 by 5-inch loaf pan. Bake at 350° for 1 hour. Cool 15 minutes before removing from pan. Makes 1 loaf.

Blue-Cran Muffin Bread

Cranberries   1 cup, chopped
Blueberries   1 cup
Sugar   ¼ cup
Flour   3 cups
Sugar   ¼ cup
Baking powder   3 teaspoons
Salt   1 teaspoon
Egg   1, beaten
Milk   1 cup
Margarine   2 tablespoons, melted
Orange peel   2 tablespoons grated or
    Dried orange peel   2 tablespoons

Mix together berries and ¼ cup sugar and set aside. Sift flour, ¼ cup sugar, baking powder, and salt into a large bowl. Combine egg, milk, and margarine and stir into flour mixture. Stir in orange peel, then berry mixture. Pour into a greased 9 by 5-inch loaf pan and bake at 350° for 1 hour. Makes 1 loaf.

Cranberry-Banana Bread

Margarine ⅓ cup
Sugar ⅔ cup
Eggs 2
Flour 2 cups
Soda 1 teaspoon
Salt ¼ teaspoon
Sour milk 3 tablespoons
Pecans ½ cup chopped
Bananas 2, mashed
Cranberries 1¼ cups, chopped

Cream margarine and sugar and add eggs 1 at a time, beating after each addition. Sift together flour, soda, and salt and add alternately with sour milk to creamed mixture. Gently stir in pecans, bananas, and berries. Pour into a greased 9 by 5-inch loaf pan and bake at 350° for 1 hour. Let cool 10 minutes, then remove from pan and cool on a rack. Makes 1 loaf.

Cranberry-Lemon Bread

Flour 1¾ cups plus 2 tablespoons
Baking powder 2½ teaspoons
Salt 1 teaspoon
Sugar ¾ cup
Butter ¼ cup
Eggs 2, beaten
Lemon peel of 1 lemon, grated
Milk ¾ cup
Flour 2 tablespoons
Cranberries ¾ cup
Lemon juice 2 tablespoons
Powdered sugar 4 tablespoons

Sift flour with baking powder and salt and set aside. Cream sugar and butter; add eggs and lemon peel and beat well. Alternately stir milk and flour mixture into creamed mixture. Mix 2 tablespoons of flour with berries; gently fold into batter. Pour into a greased 9 by 5-inch loaf pan and bake at 350° for 1 hour. Cool 10 minutes, then poke top with a fork several times. Mix lemon juice with sugar and brush over loaf. When glaze is set, remove bread from pan and serve. Makes 1 loaf.

Note This freezes very well.

Blueberry Muffins

Blueberries 1 cup
Flour ½ cup
Flour 1 cup
Salt ¼ teaspoon •
Baking powder 3 teaspoons
Margarine 2 tablespoons, softened
Sugar 2 tablespoons
Egg 1, beaten
Milk ¾ cup

Combine berries and ½ cup flour in a small bowl and toss lightly. Sift together 1 cup flour, salt, and baking powder and set aside. Cream margarine and sugar; add egg and beat well. Alternately add dry ingredients and milk and beat until just mixed. Fold in berries. Fill greased or paper-lined muffin tins ½ full and bake at 400° for 20 to 25 minutes. Makes 12 muffins.

Blueberry Nut Muffins

Eggs 2
Sugar ¾ cup
Flour 2 cups, sifted
Wheat germ ¼ cup
Baking powder 4 teaspoons
Salt ½ teaspoon
Margarine ¼ cup, melted
Milk 1 cup
Blueberries 1 cup
Flour 2 tablespoons
Nuts ½ cup chopped

Beat together eggs and sugar in a large bowl. Add flour, wheat germ, baking powder, and salt. Stir in margarine and milk. Mix together berries, flour, and nuts and add to muffin mixture. Fill greased or paper-lined muffin tins ½ full and bake at 375° for 20 to 25 minutes. Makes 16 to 18 muffins.

Deluxe Blueberry Muffins

Margarine   ¼ cup
Sugar   ¾ cup
Eggs   2
Flour   1¼ cups plus 2 tablespoons, sifted
Soda   ½ teaspoon
Salt   ¼ teaspoon
Lemon peel   1 teaspoon grated or
    Dried lemon peel   1 teaspoon
Sour cream   ¾ cup
Vanilla   ½ teaspoon
Blueberries   1 cup

Cream margarine with sugar in a large bowl. Add eggs 1 at a time, beating after each addition. Sift together flour, soda, and salt and add lemon peel. Alternately beat sifted ingredients and sour cream into creamed mixture. Add vanilla and fold in berries. Pour into greased or paper-lined muffin tins ½ full and bake at 450° for 15 minutes. Makes 12 muffins.

Cranberry Morning Muffins

Sugar   1 cup
Margarine   ¼ cup
Milk   1 cup
Egg   1
Flour   1⅓ cups
Baking powder   2 teaspoons
Cinnamon   ¾ teaspoon
Nutmeg   ¾ teaspoon
Vanilla   ½ teaspoon
Orange peel   1 teaspoon grated or
    Dried orange peel   1 teaspoon
Salt   ¼ teaspoon
Flour   ½ cup
Cranberries*   1 cup

Cream sugar and margarine in a medium bowl. Add next 9 ingredients and mix until just blended. Gently mix in remaining flour and berries. Fill greased and floured muffin tins ½ full and bake at 375° for 12 to 15 minutes. Makes 12 muffins.
*If using frozen berries add 12 minutes to cooking time.

Salads, Salad Dressings, Side Dishes, and Stuffings

Frozen Salad

Cream cheese   1 8-ounce package
Crushed pineapple   1 20-ounce can, drained
Maraschino cherries   8, chopped
Blueberries   1 cup
Whipping cream   1 pint, whipped
Miniature marshmallows   5 cups

Soften cream cheese with a fork. Add pineapple, cherries, and blueberries, stirring well between each addition. Fold in whipped cream; add marshmallows and stir until blended. Spread in a 9 by 13-inch pan and freeze. Serves 12.

Sour Cream Fruit Salad

Fresh pineapple   1 cup cut in chunks or
    Pineapple chunks   1 8-ounce can, drained
Mandarin orange segments   1 11-ounce can, drained
Blueberries   1 cup
Shredded coconut   1 cup
Sour cream   1 cup

Combine pineapple, oranges, and berries. Stir in coconut and sour cream. Refrigerate overnight or at least 8 hours. Serves 6.

Cranberry-Pineapple Mold

Crushed pineapple   1 20-ounce can, juice drained and reserved
Cherry-flavored gelatin   1 6-ounce package
Sugar   ¾ cup
Boiling water   2 cups
Cold water   ½ cup
Lemon juice   1 tablespoon
Cranberries   1 pound, ground
Small orange   1, peeled and chopped
Celery   1 cup chopped
Walnuts   ½ cup chopped

In a large bowl, stir gelatin and sugar in boiling water until dissolved. Add cold water, lemon juice, and reserved pineapple juice. Refrigerate until partially set. Add pineapple, berries, orange, celery, and walnuts. Pour into an 8-cup mold and chill until firm. Serves 12.

Sweet Cranberry Salad

Cranberries   1 pound
Sugar   2 cups
Whipping cream   1 cup, whipped
Miniature marshmallows   1 cup
Walnuts   ½ cup finely chopped

Grind berries in a food processor or grinder and mix with sugar; let stand at room temperature for 4 hours, stirring occasionally. Fold into whipped cream with remaining ingredients. Pour into an 8 by 12-inch pan and freeze until firm. Serves 12.

Raspberry-Melon Cup

Raspberries   2 cups
Sugar   to taste
Cantaloupes   2
Watermelon   2 cups seeded and diced

Purée berries in a food processor or press through a strainer with the back of a spoon; add sugar. Refrigerate until chilled. Cut cantaloupes in half crosswise; remove seeds. Remove all flesh from cantaloupe halves with melon baller and combine with watermelon chunks. Spoon melon mixture into cantaloupe halves, pour raspberry purée over each half, and serve. Serves 4.

Molded Raspberry Salad

Raspberry-flavored gelatin   1 6-ounce package
Boiling water   2 cups
Raspberries   2 cups
Frozen orange juice concentrate   1 6-ounce can, thawed
Pineapple chunks   1 20-ounce can, drained

Dissolve gelatin in boiling water, stirring constantly. Mix remaining ingredients into gelatin and pour into an 8-cup mold or large bowl. Refrigerate until set. Serves 10.

Lois's Sweetheart Salad

Cream cheese   2 3-ounce packages
Mayonnaise   1 cup
Large marshmallows   15, cut into small pieces
Strawberries   2 cups, hulled and sliced
Pecans   ½ cup chopped
Crushed pineapple   1 8-ounce can, drained
Red food coloring   ⅛ teaspoon
Whipping cream   1 cup, whipped

Beat cream cheese and mayonnaise together in a large bowl. Add next 5 ingredients and blend well. Fold in whipped cream. Pour into a 4-cup mold and chill until firm. Unmold on a bed of lettuce. Serves 8.

Lemon–Sour Cream Dressing

Sour cream   1 cup
Lemon juice   3 tablespoons
Sugar   2 tablespoons
Lemon peel   1½ teaspoons grated or
     Dried lemon peel   1½ teaspoons
Blueberries   1 cup

Combine sour cream, lemon juice, sugar, and lemon peel and mix well. Fold in berries. Use as a dressing for your favorite fruit salad combination. Makes 1½ cups.

Marinated Green Beans

Green beans   1½ pounds
Boiling salted water
Olive oil   7 tablespoons
Raspberry Vinegar   3 tablespoons
Salt   to taste
Freshly ground black pepper   to taste

Snap ends of green beans and blanch in boiling salted water until beans are barely tender; drain in a colander. Run cold water over beans and let drain again. Pour into a shallow dish. Combine olive oil, vinegar, salt, and pepper in a small bowl and mix well. Pour over beans. Chill at least 2 hours, stirring occasionally. Serves 4 to 6.

Raspberry Vinegar

Raspberries   ¾ pound
Sugar   ¼ cup
Red wine vinegar   1 12-ounce bottle

Mash berries with sugar and divide into 2 12-ounce bottles; add vinegar to fill each bottle. Set bottles uncovered on a small rack inside a large saucepan, pour enough water in saucepan to come halfway up bottles. Bring water to a boil over high heat. Lower heat and simmer for 10 minutes. Remove bottles from water and let cool. Close bottles and shake well. Makes 2 12-ounce bottles.
Note This is often called for as an ingredient in French "nouvelle cuisine."

Cranberry Nut Stuffing

Cranberries   3 cups, chopped
Dry bread cubes   3 quarts
Raisins   1½ cups
Orange peel   2 tablespoons grated
Salt   1 tablespoon
Cinnamon   ¾ teaspoon
Walnuts   ¾ cup chopped
Margarine   ¾ cup, melted
Chicken broth   1¼ cups

Combine first 7 ingredients in a large bowl and mix well, Add margarine and broth and toss lightly. Makes enough to stuff a 10- to 12-pound turkey.

Pork-Cranberry Stuffing

Bulk pork sausage   ¾ pound
Cranberries   1 cup
Celery   1½ cups chopped
Onion   ¾ cup chopped
Brown rice   3 cups cooked
Salt   1 teaspoon
Sage   ½ teaspoon
Pepper   to taste

Brown pork sausage; drain off all fat except 3 tablespoons. Put in a large mixing bowl, add remaining ingredients, and mix well. Makes enough to stuff a 4½-pound chicken.

Main Dishes and Main Dish Salads

Pork Chops with Blueberry Sauce

Thick pork chops   6
Flour   for dredging, seasoned with salt and pepper
Oil   4 tablespoons
Chicken stock   1¼ cups
Frozen orange juice concentrate   2 tablespoons, thawed
Sherry   3 tablespoons
Cornstarch   2 tablespoons
Water   2 tablespoons
Blueberries   2 cups
Cream   ¼ cup

Dredge pork chops in seasoned flour. Heat oil in a large frying pan and brown the chops. Bake at 350° for 1 hour in a covered, greased baking dish. When pork chops are almost done, heat chicken stock in a medium saucepan until boiling; add orange juice concentrate and sherry. Mix cornstarch with water and add to stock mixture; cook and stir until thickened. Reduce heat to low and stir in berries and cream. Arrange pork chops on a heated platter and pour enough sauce over to moisten. Serve the remaining sauce separately in a gravy bowl. Serves 6.

California Chicken

Whole chicken   4 to 5 pounds
Lemon juice   ⅓ cup
White wine   ⅓ cup
Oil   ⅓ cup
Salt   1 teaspoon
Oregano   1 teaspoon
Tarragon   ½ teaspoon
Rosemary   ½ teaspoon
Pepper   ½ teaspoon
Garlic cloves   3, minced
Lemon   1, quartered
Small onion   1, peeled and quartered
Oranges   2, peeled and thinly sliced
Guava   1, peeled and sliced into eighths
Fresh strawberries   1½ cups, hulled

Place chicken in a glass baking dish or large bowl; mix together next 9 ingredients and pour over chicken. Cover and chill for at least 2 hours, turning chicken several times. Drain off marinade and reserve. Insert lemon and onion into chicken cavity. Put chicken in roasting pan and bake at 350° for 1 hour, or until done, basting several times with reserved marinade. Remove chicken from oven and place on a serving platter. Arrange oranges, guava, and berries around chicken and serve. Serves 4.

Strawberry–Sour Cream Sole

Sole fillets   1 pound
Sour cream   1 cup
Tarragon   ¼ teaspoon
Chives   1 tablespoon chopped
Paprika
Strawberries   2 cups, hulled

Fold sole fillets lengthwise and arrange in a baking dish. Mix together sour cream, tarragon, and chives; spread over fish. Sprinkle each fillet with paprika. Bake at 400° for 20 to 25 minutes, depending on thickness. Arrange on a preheated serving platter and surround with berries. Serves 4.

Raspberry-Chicken Salad

Raspberries   1½ cups
Water
Sugar   ⅓ cup
Unflavored gelatin   1 ¼-ounce envelope
Cold water   ½ cup
Lemon juice   3 tablespoons
Pineapple chunks   1 15¾-ounce can, drained
Cooked chicken   1½ cups cubed
Celery   ½ cup diced
Mayonnaise   ⅓ cup
Almonds   2 tablespoons chopped
Salt   ¼ teaspoon

Push berries through a wire strainer; measure and add enough water to make 1 cup. Combine sugar and gelatin in a medium saucepan. Add raspberry mixture and heat without boiling until gelatin dissolves, stirring constantly. Remove from heat and add cold water and lemon juice. Chill until syrupy. Add pineapple and pour into a 3½-cup ring mold. Chill until firm. Meanwhile, combine remaining ingredients, stirring lightly. When raspberry mixture is set, unmold and pile chicken mixture in center. Serves 4.

Chicken-Fruit Salad

Olive oil   1 cup
Raspberry Vinegar (see Index)   ⅓ cup
Sugar   ¼ cup
French mustard   1 teaspoon
Salt   1 teaspoon
Dried mint   ¼ teaspoon
Spinach   2 bunches, washed and torn
Strawberries   2 cups, hulled and sliced
Oranges   2, peeled and chopped
Avocados   2, peeled and diced
Chicken   2 cups diced

Combine first 6 ingredients and mix well; chill. In a large salad bowl, combine remaining ingredients and toss. Pour dressing over salad and toss again. Serves 10.

Pies,
Cakes,
Frostings,
and
Fillings

Blackberry-Apple Pie

Shortcrust Pastry   for 8-inch double-crust pie
Blackberries   2 cups
Apples   2 cups peeled, cored, and chopped
Sugar   1 cup
Salt   a dash
Flour   4 tablespoons

Roll out ½ of pastry and line an 8-inch pie tin. Prick bottom of crust with a fork. Pour in berries and apples. Cover with sugar, salt, and flour. Roll out remaining pastry and cover pie; flute edges and make about 6 slits in top of pie. Bake at 400° for 10 minutes; reduce heat to 375° and bake for 20 to 30 minutes more, or until browned and bubbly. Cool before serving.

Note This is a good recipe to use when blackberries are in short supply, because the apples take on the blackberry flavor.

Shortcrust Pastry

Flour   1¾ cups
Salt   a pinch
Margarine   ½ cup
Lard* or shortening   3 tablespoons
Water   4 to 5 tablespoons

Sift flour with salt into mixing bowl. Cut in margarine and lard until mixture resembles fine bread crumbs. Add water and mix to a firm dough, then turn onto a floured board and knead lightly until smooth; do not overknead dough or it will be tough instead of flaky. Chill 15 to 20 minutes before rolling out (chilled dough is easier to roll out). Arrange dough in pie tin; pierce the bottom of the pie shell, fill, and bake according to directions. For a baked pie shell, line pie tin with pastry, pierce the bottom of the pie shell, and flute edges. Line pastry with 2 layers of tissue paper larger than the pie shell. Completely fill pie shell with dried beans and bake at 375° for 15 minutes. Remove beans by lifting out tissue paper; discard. Bake pie shell 5 minutes more; cool before filling. Makes enough for 1 8-inch double-crust pie or 2 8-inch pie shells.

Note To make pastry in a food processor, sift flour with salt and add to work bowl. Cut margarine into about 8 pieces, cut lard into small pieces, and add both to work bowl. Turn machine on and off about 12 times, or until mixture looks like bread crumbs. Add water and process until a ball of dough is formed. Turn onto a smooth, floured sur-

face and knead 3 or 4 times. Wrap and chill.
*Although some people prefer shortening, I feel lard produces much better results.

Blackberry Cream Pie

Blackberries 1 quart
Sugar 1½ cups
Soda ½ teaspoon
Salt a dash
Butter 2 tablespoons
Cornstarch 3 tablespoons
Water ¼ cup
Shortcrust Pastry 1 8-inch pie shell, baked
Whipping cream 1 cup, whipped until stiff

Combine first 5 ingredients in a medium saucepan and cook over medium heat until mixture boils. Combine cornstarch and water and add to berry mixture; continue cooking until very thick. Pour into pie shell and chill. Before serving, cover pie with whipped cream.

Blackberry-Lime Pie

Eggs 3
Lime peel of 1 lime, grated
Sugar ¾ cup
Sour cream ¾ cup
Blackberries 1½ cups
Shortcrust Pastry 1 8-inch pie shell, unbaked

Beat eggs in a medium bowl. Add lime peel and slowly beat in sugar and sour cream. Fold in berries; pour mixture into pie shell. Bake at 325° for 1 hour. Cool to room temperature and then chill.

Blackberry, Loganberry, or Boysenberry Pie

Shortcrust Pastry (see Index) for 8-inch double-crust pie
Blackberries, loganberries, or boysenberries 3½ cups
Flour 3 tablespoons
Sugar 1 cup
Cinnamon ¼ teaspoon

Roll out ½ of pastry and line an 8-inch pie tin. Pour in berries and sprinkle with flour, sugar, and cinnamon. Roll out remaining pastry and cover pie; flute edges, and make about 4 slits in top of pie. Bake at 425° for 10 minutes; reduce heat to 375° and continue baking for 25 to 30 minutes more, or until browned and bubbly. Cool before serving.

Boysenberry Cream Pie

Boysenberries 2 cups
Sugar 2 tablespoons
Boysenberries 2 cups
Sugar ¾ cup
Cornstarch 3 heaping tablespoons
Lemon juice a dash
Granola Piecrust 1 8-inch pie shell, baked
Whipping cream 1 cup, whipped

Place 2 cups of berries in a bowl and sprinkle 2 tablespoons sugar over them; set aside. Place remaining berries in a pan and add remaining sugar; mash berries well with the back of a spoon and stir until sugar begins to dissolve. Add cornstarch. Cook over medium heat until mixture is thick and clear. Cool slightly and add lemon juice. Put uncooked, reserved berries in the bottom of pie shell and pour cooked mixture over them. Chill until set. Cover with whipped cream and serve.

Granola Piecrust

Butter ½ cup, melted
Granola 2 cups, chopped
Coconut 1 cup
Oats ⅓ cup

Combine all ingredients and mix very well. Press into an 8-inch pie tin and bake at 400° for 10 minutes. Makes 1 8-inch pie shell.

Blueberry Pie

Shortcrust Pastry (see Index)   for 8-inch double-crust pie
Blueberries   3½ cups
Sugar   1 cup
Flour   ¼ cup
Salt   a dash
Lemon juice   2 teaspoons

Roll out ½ of pastry and line an 8-inch pie tin. Combine blueberries, sugar, flour, salt, and lemon juice in a bowl and mix gently. Pour into pie shell. Roll out remaining pastry and cover pie, crimping edges. Make 6 slits in top of pie and bake at 425° for 10 minutes. Reduce heat to 375° and bake for 40 minutes. Cool before serving.

Blue Lemon Cream Pie

Egg yolks   3
Sweetened condensed milk   1 14-ounce can
Lemon juice   ½ cup
Blueberries   1 cup
Granola Piecrust   1 8-inch pie shell, baked
Whipping cream   1 cup
Sugar   2 tablespoons

Combine egg yolks, milk, and lemon juice in a large bowl; stir well. Fold in berries, reserving a few for garnish. Pour into pie shell and chill 6 hours. Whip cream, adding sugar slowly; when stiff, spread over pie. Garnish edges with reserved blueberries.

Three Fruit Pie

Whole Wheat Pastry   for 9-inch double-crust pie
Blueberries   1½ cups
Apples   1 cup peeled, cored, and chopped
Crushed pineapple   ¼ cup drained
Sugar   1 cup
Salt   ⅛ teaspoon
Ground ginger   a dash
Quick-cooking tapioca   1½ tablespoons

Roll out ½ of pastry and line a 9-inch pie tin; prick bottom well. Combine berries, apples, pineapple, sugar, salt, and ginger. Let stand 1 hour. Add tapioca and pour into pie shell. Roll out remaining pastry and cover pie; crimp edges and cut 6 slits in top. Bake at 450° for 10 minutes, then reduce heat to 350° and continue cooking for 30 to 35 minutes more. Cool before serving.

Whole Wheat Pastry

Whole wheat flour   1 cup
Unbleached white flour   1 cup
Salt   1 teaspoon
Margarine   ⅔ cup
Shortening   3 tablespoons
Cold water   ¼ cup

Mix flours and salt in a large mixing bowl. Cut in margarine and shortening until mixture resembles small peas. Add water and mix just until pastry comes away from sides of bowl. Turn onto a floured surface and knead 3 or 4 times. Cover and chill at least 30 minutes before using. For double-crust pies, bake as directed in each recipe. To make baked pie shells, roll out ½ of pastry at a time and line a 9-inch pie tin. Chill for 15 minutes. Line pastry with 2 layers of tissue paper larger than the pie shell and fill pie shell with dried beans. Bake at 425°for 12 minutes, or until browned. Remove beans by lifting out tissue paper; discard. Cool pie shell before filling. Makes enough for 1 9-inch double-crust pie or 2 9-inch pie shells.
Note To make pastry in a food processor, put flours and salt in processor work bowl. Cut margarine and shortening into about 10 pieces each and add to bowl. Turn machine on and off about 8 or 9 times, or until mixture resembles small peas. Add water and process until pastry forms a ball. Remove from machine and knead a couple of times. Cover and chill at least 30 minutes before using.

Cranberry-Apple Pie

Whole Wheat Pastry   for 9-inch double-crust pie
Medium apples   3, peeled, cored, and chopped
Cinnamon   ½ teaspoon
Ground cloves   ¼ teaspoon
Brown sugar   ½ cup
Quick-cooking tapioca   3 tablespoons
Cran-Orange Sauce (see Index)   2 cups
Almonds   ½ cup chopped

Roll out ½ of pastry and line a 9-inch pie tin. Combine apples, cinnamon, cloves, and brown sugar; let stand 10 minutes. Stir tapioca,
sauce, and almonds into apple mixture. Pour into pie shell. Roll out
remaining pastry and cover pie; flute edges and cut about 8 slits in
top. Bake at 375° for about 40 minutes, or until crust is golden
brown. Serve warm or cold.

Cranberry-Raisin Pie

Whole Wheat Pastry   for 9-inch double-crust pie
Cranberries   1 cup
Sugar   ½ cup
Brown sugar   ½ cup
Raisins   ½ cup
Flour   3 tablespoons
Salt   ½ teaspoon
Orange peel   1½ teaspoons grated or
    Dried orange peel   1½ teaspoons
Medium apples   4, peeled, cored, and sliced
Margarine   2 tablespoons

Roll out ½ of pastry and line a 9-inch pie tin. Combine berries, sugar,
brown sugar, raisins, flour, salt, orange peel, and apples in a large
bowl. Toss well. Mound mixture in pie shell and dot with margarine.
Roll out remaining pastry and cover pie, crimping edges. Cut about 6
slits in top of pie. Bake at 425°for 10 minutes, then reduce heat to
350° and bake 30 to 40 minutes longer or until apples are tender and
pie is browned. Cool before serving.

Gooseberry Pie

Shortcrust Pastry (see Index)   for 8-inch double-crust pie
Gooseberries   1 quart
Sugar   1 cup
Flour   4 tablespoons
Cinnamon   ½ teaspoon
Butter   1 tablespoon

Roll out ½ of pastry and line an 8-inch pie tin. Spread berries over pastry. Combine sugar, flour, and cinnamon and sprinkle over berries. Dot with butter. Roll out remaining pastry and cover pie, crimping edges. Cut 6 slits in top of pie and bake at 400° for 45 minutes. Cool before serving.

Huckleberry Pie

Shortcrust Pastry (see Index)   for 8-inch double-crust pie
Huckleberries   3½ cups
Quick-cooking tapioca   2 tablespoons
Flour   ¼ cup
Sugar   1¼ cups*
Lemon juice   2 teaspoons
Ground cloves   a dash

Roll out ½ of pastry and line an 8-inch pie tin. Spread berries over pastry; sprinkle tapioca, flour, sugar, lemon juice, and cloves over berries. Roll out remaining pastry and cover pie; crimp edges and make 6 slits in top of pie. Bake at 425° for 10 minutes; reduce temperature to 375° and continue baking for 35 to 40 minutes more, or until filling is bubbly. Cool before serving.
* Use more sugar if berries are very tart.

Huckleberry Cream Pie

Sugar ½ cup
Cornstarch 2 tablespoons
Water ½ cup
Huckleberries 3 cups
Orange peel ½ teaspoon grated or
    Dried orange peel ½ teaspoon
Whole Wheat Pastry (see Index) 1 9-inch pie shell, baked
Orange 1
Sugar cubes 3
Whipping cream 1 cup

Combine sugar and cornstarch in a medium saucepan; stir in water and add berries. Cook over medium heat until mixture thickens, stirring constantly. Stir in orange peel. Cover and refrigerate; when cold, pour into pie shell. In a small bowl, rub sugar cubes over orange and stir until syrupy. Whip cream until it forms soft peaks, then add syrup and continue whipping until stiff. Cover pie with whipped cream and chill at least 1 hour.

Cream Cheese Raspberry Pie

Lemon juice ¼ cup
Sweetened condensed milk 1 14-ounce can
Egg yolks 2
Cream cheese 1 3-ounce package
Raspberries 1 cup
Shortcrust Pastry (see Index) 1 8-inch pie shell, baked

In a medium bowl, slowly stir lemon juice into condensed milk. In another bowl, stir egg yolks into cream cheese 1 at a time; fold into milk mixture. Fold in berries. Pour into pie shell and chill for at least 3 hours.

Deep-Dish Raspberry Pie

Shortcrust Pastry (see Index)   enough for 8-inch pie shell
Flour   3 tablespoons
Salt   ⅛ teaspoon
Lemon juice   ½ teaspoon
Sugar   ¾ cup
Raspberries   1 quart
Butter   1 tablespoon

Roll out pastry into a rectangle ½ inch wider and ½ inch longer than a 1½-quart baking dish, then set aside. In a large bowl, pour flour, salt, lemon juice, and sugar over berries; toss very gently to coat berries. Spread mixture into baking dish and dot with butter. Cover with pastry; tuck edges under and secure pastry to sides of dish by pressing with the tines of a fork. Make about 4 slits in top of pie and bake at 375° for 40 minutes, or until browned and bubbly. Cool before serving.

Raspberry Chiffon Pie

Raspberries   2 cups
Unflavored gelatin   1 ¼-ounce envelope
Cold water   2 tablespoons
Egg yolks   4, beaten
Sugar   ½ cup
Lemon juice   ½ teaspoon
Salt   ½ teaspoon
Egg whites   4
Sugar   ¼ cup
Graham Cracker Crust   1 9-inch pie shell, baked

Push berries through a wire strainer with the back of a spoon to remove seeds; set aside. Combine gelatin and water in a small saucepan and set aside. Combine egg yolks, sugar, lemon juice, and salt in top of a double boiler; beat constantly with electric mixer over boiling water until thickened. Remove from heat. Heat gelatin until dissolved and add to egg mixture. Add berries and beat for 1 minute. Chill until slightly thickened. Beat egg whites until frothy; gradually add sugar and continue beating until stiff. Fold into berry mixture. Pour into pie shell and chill until completely set.

Graham Cracker Crust

Graham cracker crumbs 1¾ cups
Sugar ¼ cup
Margarine ½ cup, melted

Combine all ingredients in a small bowl and mix well. Press into 9-inch pie tin and bake at 350° for 10 minutes. Makes 1 9-inch pie shell.

Sour Cream Raspberry Pie

Sour cream 1½ cups
Egg yolks 2, beaten
Flour ½ cup
Orange peel 1 teaspoon grated or
    Dried orange peel 1 teaspoon
Baking soda 1 teaspoon
Sugar ¾ cup
Raspberries 2 cups
Margarine 2 tablespoons
Red food coloring 8 drops
Lemon juice 1 tablespoon
Shortcrust Pastry (see Index) 1 8-inch pie shell, baked
Egg whites 2
Sugar 4 tablespoons
Vanilla ½ teaspoon
Salt ⅛ teaspoon

Heat sour cream in a double boiler until hot. In a small bowl, combine next 5 ingredients; mix well and add a little of the hot sour cream. Add this mixture to sour cream in double boiler. Stir mixture over boiling water for 10 minutes, or until thick. Add berries, margarine, and food coloring; cook another 5 minutes. Cool to room temperature and add lemon juice. Pour into pie shell. Beat egg whites until foamy, then add sugar 1 tablespoon at a time, beating well after each addition. Beat until very stiff, then add vanilla and salt. Spread over pie. Bake at 325° for 20 minutes; chill.

Yogurt Melba Pie

Raspberry yogurt   1 8-ounce container
Raspberries   1 cup
Strained peaches   2 4½-ounce jars
Frozen non-dairy whipped topping   1 8-ounce container, thawed
Graham Cracker Crust (see Index)   1 9-inch pie shell, baked

Combine yogurt, berries, and peaches in a medium bowl and mix well. Fold in whipped topping and pour into pie shell. Chill well and serve.

Glazed Strawberry Pie

Strawberries   1 cup, hulled
Water
Cornstarch   2 tablespoons
Sugar   ¾ cup
Lemon juice   ½ teaspoon
Butter   1 teaspoon
Nutmeg   a dash
Strawberries   3 cups, hulled
Graham Cracker Crust (see Index)   1 9-inch pie shell, baked
Whipping cream   1 cup, whipped until stiff

Crush 1 cup of berries and add enough water to make ½ cup; add cornstarch and sugar and cook over medium high heat until thick, stirring constantly. Remove from heat. Add lemon juice, butter, and nutmeg and chill. Place remaining berries in pie shell and cover with cooked mixture. Chill for at least 2 hours. Before serving, spread whipped cream over pie.

Spring Strawberry Pie

Milk   1½ cups
Flour   3 tablespoons
Sugar   2 tablespoons
Egg   1
Sugar   2 tablespoons
Salt   ¼ teaspoon
Vanilla   ½ teaspoon
Whole Wheat Pastry (see Index)   1 9-inch pie shell, baked
Strawberry-flavored gelatin   1 3-ounce package
Boiling water   1½ cups
Strawberries   1½ cups hulled and sliced

Scald milk. Combine flour and 2 tablespoons sugar in a medium saucepan; slowly stir in scalded milk. Cook over medium heat until mixture begins to thicken. Combine egg and remaining sugar and add to milk mixture. Continue cooking until very thick, stirring constantly. Remove from heat and add salt and vanilla. Pour into pie shell and chill in refrigerator until firm. Meanwhile, dissolve gelatin in boiling water; chill in refrigerator until it is beginning to set. Add berries. Pour over custard and chill until firm.

Red, White, and Blue Pie

Blueberries   1 cup
Sugar   ¼ cup
Water   2 tablespoons
Salt   a dash
Cornstarch   2½ teaspoons
Water   2 tablespoons
Lemon juice   1½ teaspoons
Cream cheese   1 8-ounce package
Milk   ¼ cup
Graham Cracker Crust (see Index)   1 9-inch pie shell, baked
Strawberries   1 cup hulled and sliced
Sugar   ½ cup
Salt   a dash
Water   4 tablespoons
Cornstarch   5 teaspoons
Water   4 tablespoons
Lemon juice   1 tablespoon
Red food coloring   1 drop
Fresh strawberries   3 cups, hulled

Combine blueberries, sugar, 2 tablespoons water, and salt in a small saucepan. Bring to a boil; boil 3 minutes, stirring constantly, then remove from heat. Mix cornstarch and 2 tablespoons water and stir into berry mixture; cook over medium heat until thick. Add lemon juice and cool to room temperature. Beat cream cheese with milk until smooth. Spread in pie shell; pour blueberry mixture over cream cheese and refrigerate until set. Combine sliced strawberries, sugar, salt, and 4 tablespoons water in a small saucepan. Bring to a boil; boil 3 minutes, stirring constantly, then remove from heat. Mix cornstarch and 4 tablespoons water and stir into strawberry mixture; cook over medium heat until thick. Add lemon juice and food coloring; cool to room temperature. Arrange whole strawberries, stem side down, over blueberry mixture. Brush strawberry glaze over top of pie and chill at least 2 hours.

Strawberry Chiffon Pie

Strawberries   2 cups, hulled
Sugar   ½ cup
Unflavored gelatin   1 tablespoon
Cold water   ¼ cup
Boiling water   ½ cup
Evaporated milk   1 cup, chilled
Lemon juice   1 tablespoon
Graham Cracker Crust (see Index)   1 9-inch pie shell, baked

Mash berries with sugar in a large bowl. Combine gelatin and cold water and let stand 5 minutes; add boiling water. Combine with mashed berries and stir well. Chill mixture until almost set. Combine milk with lemon juice and whip until stiff; fold into berry mixture. Pour into pie shell and chill 1 hour more.

Boysenberry Cake

Eggs   4
Sugar   1 cup
Margarine   10 tablespoons
Flour   ½ cup
Cornstarch   ½ cup
Baking powder   1 teaspoon
Ground ginger   ¼ teaspoon
Boysenberries   2 cups

In a medium bowl, beat eggs and sugar with an electric mixer until thick and light yellow. In a separate bowl, beat margarine until very soft; add to sugar and egg mixture and continue beating until fluffy. Sift together flour, cornstarch, baking powder, and ginger and carefully fold into sugar and egg mixture. Fold in berries. Pour into a greased and floured 9-inch round cake pan and bake at 375° for 25 to 30 minutes. Cool before serving.

Blueberry Cake

Flour   5 cups
Sugar   1½ cups
Baking powder   2 tablespoons
Dried orange peel   1 teaspoon
Salt   1 teaspoon
Margarine   ¾ cup
Pecans   1 cup chopped
Eggs   4
Milk   2 cups
Vanilla   2 teaspoons
Lemon juice   2 tablespoons
Blueberries   3 cups

Combine first 5 ingredients in a large bowl. Cut in margarine until mixture looks like fine crumbs; stir in pecans. Combine eggs, milk, vanilla, and lemon juice and stir into flour mixture; fold in berries. Pour into a greased and floured 10-inch bundt pan. Bake at 350° for 1½ hours, or until toothpick inserted in cake comes out clean. Cool for 10 minutes before removing from pan.

Blueberry Wallbanger Cake

Yellow cake mix   1 18½-ounce package
Instant vanilla pudding mix   1 3¾-ounce package
Oil   ½ cup
Eggs   4
Vodka   ¼ cup
Galliano   ¼ cup
Orange juice   ¾ cup
Blueberries   2 cups

Mix together first 7 ingredients and beat for 4 minutes. Fold in berries. Pour batter into a well-greased and lightly floured 10-inch bundt pan. Bake at 350° for 45 to 50 minutes. Cool 10 minutes before removing from pan.

Galles' Kansas Cake

Flour 2 cups
Soda 2 teaspoons
Salt ½ teaspoon
Eggs 2
Sugar 1½ cups
Fruit cocktail 1 16-ounce can, undrained
Blueberries 1 cup
Brown sugar ½ cup
Nuts ½ cup chopped
Shredded coconut 1 cup
Sugar ¾ cup
Evaporated milk ½ cup
Margarine ½ cup
Vanilla 1 teaspoon

Sift together flour, soda, and salt and set aside. Combine eggs and sugar and mix well. Alternately add flour mixture and fruit cocktail to egg mixture. Fold in berries and pour into a 9 by 13-inch pan. Sprinkle brown sugar and nuts over top of cake. Bake at 350° for 45 minutes. While cake is still hot, sprinkle with coconut. Cook remaining ingredients in a small saucepan over medium heat until margarine is melted, then increase heat to high and boil 1 minute. Pour over cake.

Cranberry Cake

Cranberries 2 cups
Sugar 2 tablespoons
Walnuts ½ cup chopped
Eggs 2
Sugar ½ cup
Butter ½ cup, melted
Oil ½ cup
Flour 1 cup

Place berries in a 9 by 13-inch pan. Sprinkle with sugar and walnuts and set aside. Beat eggs with electric mixer and gradually add sugar; beat until light. Combine butter and oil and add alternately with flour to creamed mixture. Spread batter over cranberries and nuts. Bake at 325° for 50 minutes.

Cranberry-Pineapple Upside-Down Cake

Margarine   ¼ cup, melted
Brown sugar   ¾ cup
Pineapple chunks   1 15¼-ounce can, drained
Cranberries   ¾ cup
Yellow cake mix   1 9-ounce package

Pour margarine into an 8 by 8-inch pan and sprinkle with brown sugar. Distribute pineapple and berries over margarine mixture. Mix cake according to package directions and pour over fruit. Bake as directed on package. Remove from oven and cool 15 minutes. Turn upside-down onto a serving plate. Serve warm.

Cranberry Spice Cake

Margarine   ¾ cup
Brown sugar   1¼ cups
Sugar   1 cup
Eggs   3
Flour   2¾ cups
Soda   1½ teaspoons
Cinnamon   1½ teaspoons
Nutmeg   ¾ teaspoon
Ground cloves   ¾ teaspoon
Ground ginger   ⅛ teaspoon
Salt   1 teaspoon
Buttermilk   1½ cups
Cranberries   1 cup, chopped

Cream margarine, brown sugar, and sugar; beat in eggs 1 at a time. Sift together next 7 ingredients; mix alternately with buttermilk into creamed mixture. Fold in berries. Pour into a greased 9 by 13-inch pan and bake at 350° for 50 minutes.

Gooseberry Upside-Down Cake

Gooseberries   1 quart
Sugar   1¾ cups
Flour   2 cups
Baking powder   3 teaspoons
Salt   ½ teaspoon

Sugar ½ cup
Cinnamon ½ teaspoon
Milk ¾ cup
Egg 1, beaten
Margarine 4 tablespoons, melted

Mix berries and sugar and spread in a greased 9 by 13-inch pan. Sift together next 5 ingredients. Combine milk, egg, and margarine and add to sifted dry ingredients; stir until well mixed. Spread over berries. Bake at 350° for 30 to 40 minutes. Cool in pan, then invert onto a serving plate.

Huckleberry-Orange Cake

Flour 2½ cups
Sugar 1 cup
Salt ½ teaspoon
Soda 1 teaspoon
Baking powder 1 teaspoon
Pitted dates 1 8-ounce package, quartered
Huckleberries 2 cups
Orange peel 2 tablespoons grated or
    Dried orange peel 2 tablespoons
Walnuts 1 cup chopped
Eggs 2, beaten
Oil ¾ cup
Milk 1 cup
Glaze (optional)

Sift first 5 ingredients into a large bowl; stir in dates, berries, orange peel, and walnuts. Combine eggs, oil, and milk and mix well; pour into flour mixture and stir until well combined. Pour into a greased and floured 10-inch bundt pan. Bake at 350° for 1 hour. Cool 10 minutes and remove from pan; spread with glaze, if desired, while still warm.

Glaze

Powdered sugar 1 cup, sifted
Water 1 tablespoon
Corn syrup 1 tablespoon
Cinnamon a dash

Mix all ingredients in a small bowl.

Pat's Rosy Cake

Flour   2 cups
Baking powder   2½ teaspoons
Salt   ¼ teaspoon
Brown sugar   ¼ cup
Margarine   ½ cup
Egg   1, beaten
Milk   ¾ cup
Rhubarb   2 cups finely diced
Strawberries   3 cups, hulled and sliced
Strawberry-flavored gelatin   1 3-ounce package
Butter   6 tablespoons
Sugar   1½ cups
Flour   ½ cup

Combine flour, baking powder, salt, and brown sugar in a large bowl; cut in margarine. Add egg and milk and mix thoroughly with electric mixer. Spread in a greased 9 by 13-inch pan, distribute rhubarb and berries on top, and sprinkle with gelatin. Mix together butter, sugar, and flour until crumbly and sprinkle over top of cake. Bake at 350° for 50 minutes.

Strawberry Cake

Unflavored gelatin   1 teaspoon
Water   1 tablespoon
Strawberries   2 cups, hulled and sliced
Sugar   2 tablespoons
Whipping cream   2 cups
Sugar   3 tablespoons
Angel food cake   1 10-inch round cake

Soften gelatin in water; melt in the top of a double boiler over hot, but not boiling, water. Mash berries with 2 tablespoons sugar; add melted gelatin and stir well. Chill until partially set. Whip cream, gradually adding remaining sugar; measure out 1 cup and reserve remainder in refrigerator. Fold the cup of whipped cream into partially set strawberry mixture and chill again. Split cake into 3 layers and spread each layer with strawberry mixture. Frost cake with remaining sweetened whipped cream. Chill for at least 1 hour before serving.

Fluffy Strawberry Frosting

Egg whites 2
Sugar ½ cup
Strawberries ½ cup hulled and sliced
Orange liqueur 2 teaspoons
Lemon juice 1 tablespoon

Combine all ingredients in the top of a double boiler and place over boiling water. Beat with an electric mixer for 6 to 7 minutes, or until mixture holds stiff peaks. Spread over cake. Makes enough to frost 2 8-inch layers.
Note This frosting is soft, but it holds well.

Blueberry-Orange Filling

Sugar 1 cup
Cake flour 5 tablespoons
Orange peel 2 tablespoons grated or
    Dried orange peel 2 tablespoons
Orange juice ½ cup
Lemon juice 3 tablespoons
Water 4 tablespoons
Egg 1, beaten
Butter 2 teaspoons
Blueberries ½ cup

Mix together all ingredients except berries in the top of a double boiler. Cook over boiling water for 10 minutes, stirring constantly. Remove from heat and cool slightly. Fold in berries. Cool to room temperature before using. Makes enough to fill 1 9-inch layer cake.

Ice Cream, Sherbet, Desserts, and Confections

Blackberry Ice Cream

Egg yolks   6
Whole milk   2 cups
Honey   1 cup
Salt   ¼ teaspoon
Blackberries   3 cups, crushed
Cream   1 pint
Lime juice   2 tablespoons

Beat egg yolks in a large bowl. Add milk and stir well, then add honey and salt. Pour mixture into the top of a double boiler and stir constantly over medium heat until it coats the back of a spoon. Turn back into bowl, cover, and chill until cold. Stir in berries, cream, and lime juice. Freeze in an ice cream freezer according to manufacturer's directions. Makes 2½ quarts.

Easy Blackberry Ice Cream

Blackberries   1 quart
Sugar   1½ cups
Water   1½ cups
Whipping cream   ½ cup, whipped to soft peaks

Combine berries, sugar, and water and purée ½ of the mixture at a time in a food processor or blender; fold in whipped cream. Chill 1 hour. Pour into a 2-quart glass or metal bowl and freeze until mixture is frozen about 2 inches from sides of bowl. Whip with an electric mixer until smooth; refreeze. Continue this procedure 2 more times and then freeze until solid. Makes 2 quarts.

Loganberry-Apple Sherbet

Sugar   ⅔ cup
Water   ⅔ cup
Loganberries   1¼ cups
Apples   3, peeled and cored
Brandy   2 tablespoons

Combine sugar and water in a small saucepan and stir until sugar is dissolved. Cook over high heat until mixture is about to boil; remove from heat, cover, and cool. Purée berries in a food processor, then press through a strainer to remove seeds. Discard seeds. Purée apples in a food processor. Combine all ingredients in a large bowl and mix well. Freeze in an ice cream freezer, following manufacturer's directions for sherbet. Makes 1 quart.

Cranberry Ice

Cranberries 2 cups
Water 4 cups
Sugar 1 cup or to taste
Frozen orange juice concentrate 3 teaspoons, thawed

Combine berries and water in a large saucepan and bring to a boil over high heat. Reduce heat and cover pan tightly; simmer for about 10 to 12 minutes. Purée berries and water in a food mill or processor. Press mixture through a strainer into a large bowl and stir in sugar and orange juice concentrate. Pour into 3 ice cube trays and freeze. To serve, remove cubes from trays and grate with a hand grater or put through a food processor with a steel blade. Spoon into dessert dishes and return each serving to freezer until you have processed the desired number of servings. Makes 1 quart.

Quick Raspberry Sherbet

Raspberries 1¼ cups
Lemon juice 1½ tablespoons
Powdered sugar ¾ cup

Purée berries in blender at medium speed by adding a few at a time to blender container. Add lemon juice and powdered sugar; blend about 1 minute longer. Pour into 4 dessert dishes, cover, and freeze for at least 1 hour. Remove from freezer 15 minutes before serving. Serves 4.

Strawberry Sherbet

Sugar ⅔ cup
Water ⅔ cup
Strawberries 5 cups, hulled
Orange juice 2 tablespoons
Lemon juice 2 tablespoons

Combine sugar and water in a small saucepan and cook over medium heat, stirring constantly, until sugar dissolves; do not boil. Remove from heat and chill in refrigerator until cold. Purée berries in a food processor or chopper. When sugar mixture is cold, combine with strawberry purée, orange juice, and lemon juice; stir well. Process in an ice cream freezer according to manufacturer's directions. Makes 1 quart.

Blackberry Bavarian Cream

Sugar 4 tablespoons
Water ⅔ cup
Unflavored gelatin 1 tablespoon
Milk 4 tablespoons
Blackberries 1½ cups
Egg yolks 2
Cream 1 cup
Crushed ice 1 cup
Whipped cream garnish (optional)

Combine sugar and water in a saucepan and bring to a boil; pour mixture into blender. Add gelatin and milk and blend at high speed for 1 minute. Add berries and egg yolks; blend for 10 seconds. Add cream and crushed ice; blend until smooth. Pour into a 5-cup mold and chill until firm. Unmold onto a serving dish and garnish with whipped cream if desired. Serves 6.

Blackberry Dumplings

Shortcrust Pastry (see Index)　enough for 8-inch double-crust pie
Blackberries　1 quart
Sugar　¾ cup
Flour　3 tablespoons
Sugar　¾ cup
Butter　2 tablespoons
Flour　1 tablespoon
Hot water　1½ cups
Ice cream　topping (optional)

Roll pastry into a rectangle and cut into 6 6-inch squares. Divide berries evenly among squares. Combine ¾ cup sugar and 3 tablespoons flour and sprinkle over berries. Seal each square by pinching edges together; place dumplings in a buttered deep baking dish. Combine remaining ingredients except ice cream and stir until sugar is dissolved and butter is melted. Pour liquid over dumplings and bake at 425° for 40 minutes, basting once. Serve warm or cold; top with ice cream if desired. Serves 6.

Blackberry Pudding

Blackberries　2 cups
Salt　¼ teaspoon
Sugar　½ cup
Cinnamon　½ teaspoon
Vanilla　½ teaspoon
Flour　1 cup
Baking powder　2 teaspoons
Salt　½ teaspoon
Sugar　¾ cup
Milk　1 cup
Butter　½ cup, melted

Combine first 5 ingredients in a greased 1-quart baking dish; toss gently. Sift together flour, baking powder, salt, and sugar and put in a large bowl. Combine milk and butter and add to dry ingredients; pour over berry mixture. Bake at 400° for 30 minutes. Serves 4 to 6.

Blackberry Trifle

Ladyfingers   12
Sherry   to taste (optional)
Blackberry-flavored gelatin   1 3-ounce package
Blackberries   1 cup
Instant vanilla pudding mix   1 3¾-ounce package
Blackberries   1 cup .
Whipping cream   1 pint
Vanilla   1 teaspoon
Slivered almonds   ¼ cup
Maraschino cherries   8, halved

Arrange ladyfingers on the bottom and sides of a large glass bowl. Soak with sherry if desired. Make gelatin as directed on package and chill until partially set. Place 1 cup of berries over ladyfingers and cover with partially set gelatin. Chill until completely set. Mix pudding according to package directions, pour over gelatin, and chill again. Cover with remaining berries. Whip cream with vanilla and spread over trifle; decorate with slivered almonds and cherry halves. Serves 10.

Boysenberry Crunch

Boysenberries   1 quart
Sugar   1 cup
Cinnamon   ½ teaspoon
Ground cloves   ¼ teaspoon
Quick-cooking tapioca   3 tablespoons
Sugar   ½ cup
Flour   ½ cup
Butter   ⅓ cup
Ice cream   topping (optional)

Combine first 5 ingredients in a 9 by 13-inch baking dish and bake at 375° until mixture is boiling. Meanwhile, mix sugar and flour and cut in butter with a fork until mixture is crumbly. Sprinkle over berry mixture and continue baking for 30 minutes more. Serve with ice cream. Serves 8.

Quick Boysenberry Dessert

Blackberry or boysenberry yogurt ½ cup
Whipped cream cheese 1 4-ounce container
Sugar ¼ cup
Boysenberries 1 cup
Pound cake slices 6

Stir together yogurt, cream cheese, and sugar until well combined; carefully fold in berries. Serve on pound cake slices. Serves 6.

Loganberry-Blackberry Pudding

Loganberries 3 cups
Blackberries 2 cups
Sugar ⅔ cup
Cornstarch 2 tablespoons
Water ¾ cup
Lemon juice 1 tablespoon
Whipping cream 1 cup
Orange flavoring ½ teaspoon
Sugar 1 tablespoon

Combine berries and sugar in a medium saucepan and bring to a boil over medium heat. Remove from heat and push through a strainer to remove seeds. Discard seeds and return berry mixture to pan; bring to a boil. Mix together cornstarch and water and add to berry mixture; cook until thick, stirring constantly. Remove from heat and add lemon juice. Pour into 8 dessert dishes and chill. Before serving, whip cream, adding orange flavoring and sugar when cream is almost stiff. Spoon onto pudding. Serves 8.

Loganberry Cobbler

Loganberries   1 quart
Sugar   ⅓ cup
Brown sugar   ⅓ cup
Lemon juice   ½ teaspoon
Butter   2 tablespoons
Biscuit mix   1½ cups
Butter   3 tablespoons, melted
Egg   1, beaten
Milk   ½ cup

In a large bowl, mix berries with sugar, brown sugar, and lemon juice. Pour into a greased 8 by 8-inch pan and dot with butter. Combine biscuit mix, melted butter, egg, and milk and mix until just moistened. Drop biscuit mixture by spoonfuls over berries. Bake at 400° for 30 minutes or until brown. Serves 6.

Loganberry-Strawberry Compote

Loganberries   2½ cups
Frozen strawberries   2 10-ounce packages, thawed, juice drained
   and reserved
Sugar   ¾ cup or to taste
Cinnamon   ½ teaspoon
Ground ginger   a dash
Ground cloves   ⅛ teaspoon
Cornstarch   2 tablespoons
Whipping cream   1 cup, whipped

Combine berries, sugar, and spices in a large saucepan and bring to a boil over medium heat. Combine ¼ cup strawberry juice and cornstarch in a small bowl and add to berries; continue cooking, stirring constantly, until slightly thickened. Cover and refrigerate until cold; pour into a serving bowl and serve. Pass whipped cream separately. Serves 6 to 8.

Blueberries Jubilee

Sugar   1 cup
Water   1½ cups
Blueberries   2 cups
Cornstarch   1 tablespoon
Water   3 tablespoons
Brandy   ¼ cup, heated*
Vanilla ice cream   1 quart, cut into 8 slices

Combine sugar and water in a medium saucepan and bring to a boil. Boil for 5 minutes, then remove from heat and add berries. Combine cornstarch and water; add to berry mixture. Cook over medium heat until thick and bubbly, stirring constantly. Pour into chafing dish or deep 12-inch frying pan, add brandy, and ignite. Serve immediately over ice cream slices. Serves 8.
* Be sure only to heat brandy; do not boil it.

Easy Blueberry-Peach Cobbler

Sugar   ⅔ cup
Cornstarch   4 teaspoons
Water   1 cup
Peaches   6, peeled and sliced
Blueberries   1 cup
Lemon juice   1 teaspoon
Refrigerated cinnamon rolls   1 9½-ounce package

Combine sugar, cornstarch, and water in a medium saucepan; bring to a boil, stirring constantly. Add peaches and berries and bring back to a boil. Cook and stir until thick. Remove from heat and add lemon juice. Pour into a shallow 2-quart baking dish. Lay cinnamon rolls evenly across top. Bake at 400° for 15 minutes. Serve warm. Serves 6 to 8.

Blueberry Buckle

Sugar   ¾ cup
Margarine   ¼ cup, softened
Egg   1, beaten
Milk   ½ cup
Flour   2 cups
Baking powder   2 teaspoons
Salt   ½ teaspoon
Cinnamon   ½ teaspoon
Blueberries   1 cup
Sugar   ½ cup
Flour   ⅓ cup
Nutmeg   ¼ teaspoon
Margarine   ¼ cup
Lemon Sauce

Cream together sugar and margarine. Add egg and milk and mix thoroughly. Sift together flour, baking powder, salt, and cinnamon, stir into creamed mixture, and fold in berries. Pour into a greased and floured 9 by 9-inch pan. Mix sugar, flour, nutmeg, and margarine in a small bowl; sprinkle over mixture in pan. Bake at 375° for 45 to 50 minutes. When cool, serve topped with Lemon Sauce. Serves 9.

Lemon Sauce

Butter   ½ cup
Sugar   1 cup
Water   1¼ cups
Egg   1, beaten
Lemon juice   3 tablespoons
Lemon peel   1 tablespoon grated or
    Dried lemon peel   1 tablespoon

Blend all ingredients together in a saucepan and bring to a boil over medium high heat. Remove from heat and serve. Makes 2 cups.

Frozen Blueberry Cream

Blueberries 2 cups
Large marshmallows 24, finely chopped
Crushed pineapple 1 8-ounce can, drained
Sugar ¼ cup
Salt ⅛ teaspoon
Whipping cream 1 cup, whipped

Combine first 5 ingredients in a large bowl and mix well. Fold in whipped cream and pour into an 8 by 8-inch pan. Freeze for at least 4 hours. Serves 8.

Blueberry Gingerbread

Molasses 1 cup
Buttermilk 1 cup
Flour 2¼ cups
Baking powder 1¾ teaspoons
Ground ginger 2 teaspoons
Salt ½ teaspoon
Egg 1, beaten
Margarine ½ cup, melted
Blueberries 1½ cups

In a large bowl, stir together molasses and buttermilk until combined. Sift together flour, baking powder, ginger, and salt; stir into molasses mixture. Add egg and margarine and stir until creamy. Fold in berries. Pour into a greased 9 by 9-inch pan. Bake at 350° for 40 minutes.
Note This is delicious served warm with whipped cream.

Blueberry Soup

Blueberries   1 quart
Water   3 cups
Sugar   ¾ cup
Cinnamon stick   1 2-inch piece
Cornstarch   2 tablespoons
Water   4 tablespoons
Sour cream   topping (optional)

Combine berries, water, sugar, and cinnamon in a medium saucepan and bring to a boil; reduce heat and cook until berries are soft. Discard cinnamon stick. Mix cornstarch and water and add to berry mixture. Bring back to a boil and stir until clear. Cover and chill. Serve in small bowls, with a spoonful of sour cream if desired. Serves 6.
Variation Add ½ cup of Burgundy wine before chilling.

Easy Blueberry Cheesecake

Cream cheese   1 8-ounce package, softened
Milk   ½ cup
Milk   1½ cups
Instant vanilla pudding mix   1 3¾-ounce package
Blueberries   1½ cups
Graham cracker crust   8- or 9-inch pie shell

Beat together cream cheese and ½ cup milk. Add the remaining milk and the pudding mix and beat for 1 minute. Chill mixture in refrigerator for about 3 minutes, or until it begins to set, then fold in berries. Pour into pie shell. Chill for at least 1 hour.

Sauced Blueberries

Blueberries   2 cups
Whipping cream   1 cup
Honey   to taste

Divide berries into 4 dessert dishes. Beat cream until just starting to thicken; blend in honey to taste. Whip cream to soft peaks and spoon onto berries. Serves 4.

Frozen Blueberry Torte

Graham cracker crumbs 1½ cups
Pecans ½ cup chopped
Sugar ¼ cup
Margarine 6 tablespoons, melted
Blueberries 2 cups
Sugar 1 cup
Egg whites 2
Lemon juice 1 tablespoon
Vanilla 1 teaspoon
Salt ⅛ teaspoon
Cream of tartar ¼ teaspoon
Whipping cream 1 cup, whipped to soft peaks

Combine crumbs, pecans, sugar, and margarine and press into bottom and sides of a 10-inch springform pan. Chill well. Combine berries and sugar in a large mixing bowl and let stand 10 minutes. Add next 5 ingredients. Beat with electric mixer at low speed until frothy, then turn speed to high and beat for 6 to 8 minutes, or until mixture forms stiff peaks. Fold whipped cream into berry mixture; turn into springform pan. Cover and freeze until firm. To serve, remove from pan and place on a serving plate. Serves 8 to 10.

Upside-Down Blueberry Cobbler

Flour 1½ cups
Baking powder 2 teaspoons
Salt a dash
Sugar 1 cup
Margarine ¼ cup, melted
Milk 1 cup
Blueberries 1 quart
Sugar ¾ cup
Boiling water 2 cups

Sift flour, baking powder, and salt into a large bowl. Add sugar and stir in margarine and milk. Spread in a 9 by 13-inch pan. Cover with berries; sprinkle sugar over berries. Pour boiling water over cobbler and bake at 350° for 50 to 60 minutes. Serves 8.

Holiday Cheesecake

Graham cracker crumbs   1¼ cups
Sugar   ¼ cup
Margarine   ¼ cup, melted
Cream cheese   2½ 8-ounce packages
Sugar   ¾ cup
Flour   4 teaspoons
Vanilla   1 teaspoon
Nutmeg   ¼ teaspoon
Cinnamon   ½ teaspoon
Milk   3 tablespoons
Egg yolks   4
Cranberries   1½ cups
Water   ½ cup
Sugar   ½ cup
Brown sugar   ¼ cup
Walnuts   ¼ cup chopped

Combine cracker crumbs, sugar, and margarine and mix well; press onto the bottom and sides of a greased 9-inch springform pan. Beat cream cheese with electric mixer until soft, then slowly add sugar. Continue beating and add next 5 ingredients. Add egg yolks 1 at a time, beating after each addition. Pour into prepared pan and bake in preheated oven at 500° for 8 minutes. Reduce heat to 200° and bake 45 minutes more. Cool, then refrigerate. Meanwhile, bring berries and water to a boil over medium high heat and continue cooking for 10 minutes. Add sugar and brown sugar and cook about 3 minutes more, or until sugar dissolves. Remove from heat; add walnuts. Chill mixture until cold; pour over cheesecake and chill again. To serve, unmold cheesecake onto a serving dish and cut into small wedges. Serves 12.

Gooseberry Dessert Soup

Gooseberries 1 cup
Water 2 tablespoons
Sugar ¼ cup
Milk 3 cups
Egg 1
Flour 1 tablespoon
Milk 1½ tablespoons
Cinnamon ⅛ teaspoon
Whipping cream ½ cup, whipped

Combine berries, water, and sugar in a medium saucepan and simmer until berries are soft. Turn heat to high and add milk, stirring constantly. Mix together egg, flour, and milk, and beat well; add a little of the hot mixture to this and combine with mixture in pan. Boil 2 minutes, stirring constantly. Remove from heat and add cinnamon; cover and chill. To serve, pour into 4 small bowls and top with whipped cream. Serves 4.

Gooseberry Fool

Cornstarch 1 teaspoon
Sugar 2 tablespoons
Milk ⅔ cup
Egg yolks 2, beaten
Vanilla ⅛ teaspoon
Gooseberry Sauce (see Index) 2 cups, cooled
Whipping cream 1 cup, whipped until stiff

Combine cornstarch and sugar in a small saucepan and slowly stir in milk; bring to a boil. Remove from heat and add a little hot mixture to egg yolks, then pour them into pan with milk mixture. Cook over low heat (do not let mixture boil) until mixture coats the back of a spoon. Remove from heat and add vanilla; chill. When cool, fold into gooseberry sauce, then fold in whipped cream. Pour into a serving bowl or 6 dessert dishes. Serves 6.

Gooseberry Supreme

Graham cracker crumbs   1½ cups
Margarine   ¼ cup, melted
Sugar   ¾ cup
Cream cheese   1 8-ounce package, softened
Eggs   2, beaten
Vanilla   1 teaspoon
Dried orange peel   ½ teaspoon
Gooseberry Sauce (see Index)   2 cups

Combine cracker crumbs, margarine, and sugar in a small bowl; blend well. Press onto the bottom and sides of a greased 8 by 8-inch pan. Mix cream cheese, eggs, vanilla, and orange peel with an electric mixer until smooth. Spread over crumb mixture and bake at 300° for 30 minutes. Remove from oven and cool to room temperature. Pour gooseberry sauce over top and chill for at least 1 hour. Serves 8.

Petry's Gooseberry Cobbler

Gooseberry Sauce (see Index)   2 cups
Flour   1½ cups
Salt   ½ teaspoon
Baking powder   3 teaspoons
Cinnamon   ½ teaspoon
Whole wheat flour   ¼ cup
Sugar   1 tablespoon
Margarine   6 tablespoons
Half-and-half   ¾ cup

Pour gooseberry sauce into the bottom of a greased 8 by 8-inch pan and set aside. Sift flour, salt, baking powder, and cinnamon into a large bowl. Add whole wheat flour and sugar. Cut in margarine until mixture looks like small peas. Add half-and-half and stir until mixture begins to leave the sides of the bowl. Turn onto a floured board and knead about 5 times. Roll dough into an 8-inch square and place on top of gooseberry sauce. Bake at 450° for 12 to 15 minutes, or until golden brown. Serves 6.

Raspberry Crunch

Raspberries 1 quart
Sugar ⅓ cup
Dried lemon peel ½ teaspoon
Margarine ¼ cup
Flour ⅓ cup
Brown sugar ⅓ cup
Oats ¾ cup

Place berries in a 9 by 9-inch pan; sprinkle with sugar and lemon peel. Cut margarine into flour, brown sugar, and oats and sprinkle over raspberries. Bake at 350° for 30 minutes. Serves 6.

Raspberry Fool

Raspberries 1½ cups
Sugar ¼ cup
Whipping cream 1 cup
Sugar 2 tablespoons
Vanilla ice cream 1 quart

Purée berries and press through a metal strainer to remove seeds. Stir in ¼ cup sugar and chill in freezer until firm but not frozen. Whip cream with 2 tablespoons sugar until soft peaks form. Gradually beat in ice cream until mixture is fluffy but ice cream is not melted. Place in freezer for about 15 minutes, or until firm. Be sure the two mixtures are approximately the same consistency, then remove both mixtures from freezer and fold together. Serve immediately. Serves 8.

Huckleberry Betty

Fresh bread cubes   4 cups cut in ½-inch cubes
Butter   ½ cup, melted
Sugar   ½ cup
Cinnamon   1 teaspoon
Lemon juice   2 tablespoons
Brown sugar   ½ cup
Huckleberries   2 cups

In a large bowl, toss together bread cubes, butter, sugar, and cinnamon until combined. In another bowl, sprinkle lemon juice and brown sugar over berries. Alternately spread layers of bread mixture and berry mixture in an 8 by 12-inch baking dish. Bake at 350° for 20 to 30 minutes, or until bubbly. Serves 8.

Lingonberry Mousse

Water   1½ cups
Salt   ¼ teaspoon
Quick-cooking farina   ¼ cup
Lingonberry preserves   1 14¾-ounce jar
Whipped cream   topping (optional)

Combine water and salt in a medium saucepan and bring to a boil. Add farina slowly and cook for 2½ minutes, stirring constantly; remove from heat and add preserves. Cool mixture to lukewarm and beat with an electric mixer until light and fluffy. Pour into 6 dessert dishes and chill. Serve with whipped cream if desired. Serves 6.

Peach Melba

Raspberries   2 cups
Sugar   ⅔ cup
Cornstarch   1 tablespoon
Lemon juice   1 tablespoon
Peaches   3, peeled and halved
Lemon juice   1 tablespoon
Vanilla ice cream   1 pint

Crush berries well, and set aside. In a medium saucepan, combine sugar and cornstarch and add half the berries. Cook over medium heat until sauce boils and thickens, stirring constantly. Add remaining berries and 1 tablespoon lemon juice. Cover and cool in refrigerator. Meanwhile, brush peach halves with lemon juice. Divide ice cream into 6 dessert dishes. Top each with a peach half and spoon 2 tablespoons of cooled raspberry sauce over each peach. Serves 6.

Raspberry-Lemon Sundaes

Raspberries   1½ cups, sweetened to taste, or
   Frozen raspberries   1 10-ounce package, thawed
Lemon sherbet   1 pint

Smash berries with a large spoon or whirl in a food processor until smooth. Make 4 large scoops out of sherbet and put into dessert dishes. Spoon raspberry purée over sherbet and serve. Serves 4.

Raspberry Pudding

Raspberries   1 quart
Orange juice   2 cups
Water   2 cups
Sugar   1½ cups
Cornstarch   ¾ cup
Water   ⅔ cup
Sugar   ½ cup
Whipped cream   topping (optional)

Combine berries, juice, and water in a large saucepan and bring to a boil, stirring constantly; cook until fruit is soft. Remove from heat and press fruit mixture through a strainer to remove seeds. Return juice to pan; add sugar and stir until sugar dissolves. Mix cornstarch with water and stir into juice. Bring to a boil over medium high heat and cook until thick, stirring constantly. Remove from heat and pour into 8 dessert dishes or a large bowl. Sprinkle with sugar; chill. Serve with whipped cream if desired. Serves 8.

Raspberry Shortbread Dessert

Butter   1 cup
Flour   2 cups
Powdered sugar   ½ cup
Cornstarch   1 tablespoon
Salt   ¼ teaspoon
Sugar   to taste
Raspberries   1 quart
Whipping cream   1 cup, whipped

Cream butter. Sift together flour, powdered sugar, cornstarch, and salt. Work flour mixture into butter with fingers and press mixture into an 8 by 12-inch pan. Mark 12 equal squares with the tines of a fork. Bake at 325° for 25 minutes. Cut through marked squares while still warm, then let cool completely. Meanwhile, add sugar to berries and chill. To serve, place equal amounts of whipped cream on each cookie and spoon berries on top of whipped cream. Serves 12.

Raspberry Supreme

Water   ¼ cup
Large marshmallows   32
Raspberries   1½ cups
Whipping cream   1 cup, whipped

Combine water and marshmallows in the top of a double boiler; heat over boiling water until melted, stirring constantly. Remove from heat and stir in berries. Chill until mixture thickens; fold in whipped cream. Pour into 6 dessert dishes and chill for at least 4 hours. Serves 6.

Raspberry Tapioca

Quick-cooking tapioca   3 tablespoons
Sugar   4 tablespoons
Salt   ¼ teaspoon
Milk   1⅓ cups
Whipping cream   ½ cup, whipped until stiff
Raspberries   2 cups
Sugar   2 tablespoons
Whipped cream   topping (optional)

In a medium saucepan, combine tapioca, sugar, and salt and gradually stir in milk. Bring to a boil over medium high heat, stirring constantly. Remove from heat and cool. Whip cream until stiff; fold into cooled tapioca mixture and chill. Crush berries and add sugar. Layer tapioca mixture and berries in 4 tall glasses or pour tapioca mixture over berries in a large bowl. Serve with whipped cream if desired. Serves 4.

Sour Cream Strawberry Parfait

Sour cream   1½ cups
Brown sugar   ½ cup
Lemon juice   2 tablespoons
Strawberries   1 quart, hulled and sliced
Mint sprigs   garnish (optional)

Mix sour cream with brown sugar until smooth. Stir in lemon juice, cover, and chill. Layer berries and sauce in 8 parfait glasses. If desired, garnish with mint sprigs. Serves 8.

Strawberry Fondue

Frozen strawberries   2 10-ounce packages, thawed
Cornstarch   ¼ cup
Sugar   2 tablespoons
Water   ½ cup
Whipped cream cheese   1 4-ounce container
Triple Sec   ¼ cup
Strawberries   for dunking
Pound cake cubes   for dunking
Pineapple chunks   for dunking
Marshmallows   for dunking
Peach slices   for dunking
Pear slices   for dunking

In a medium saucepan, crush berries slightly. Blend together cornstarch, sugar, and water; add to berries and stir well. Cook over medium heat, stirring constantly, until thickened and bubbly. Pour into fondue pot and place over fondue burner. Add cream cheese and stir until melted. Gradually add Triple Sec. Spear dunking foods with fondue forks and dip into fondue. Serves 8.

Strawberry Macaroon Parfait

Strawberries   2 cups, hulled
Sugar   2 tablespoons
Egg yolks   2
Powdered sugar   ¼ cup
Margarine   2 tablespoons, melted
Strawberry liqueur   1 tablespoon (optional)
Vanilla   1 teaspoon
Frozen whipped topping   1 cup, thawed
Macaroons   4, crumbled

Set aside 4 whole berries. Slice remaining berries, combine with sugar, and let stand 15 minutes. Meanwhile, beat egg yolks in a small bowl until very thick. Beat in powdered sugar. Stir in margarine, liqueur, and vanilla. Fold in whipped topping. Spoon ½ the berries into 4 parfait glasses and top with ½ the macaroons and ½ the egg mixture. Repeat layers. Garnish with reserved berries and chill. Serves 4.

Strawberry 'n' Cream Crepes

Strawberries   3 cups hulled and sliced
Cornstarch   4 teaspoons
Sugar   ¾ cup
Water   1 cup
Orange peel   1 teaspoon grated or
    Dried orange peel   1 teaspoon
Rum   2 tablespoons (optional)
Strawberries   1 cup hulled and sliced
Whipping cream   1½ cups
Powdered sugar   to taste
Dessert Crepes   14 to 16

Purée berries. Combine cornstarch and sugar in a medium saucepan, add puréed berries and water, and stir well. Boil for 1 minute; add orange peel and rum. Cool. Stir in remaining berries, cover, and chill for at least 2 hours. Whip cream to soft peaks, adding powdered sugar to taste. Spread each crepe with about 3 tablespoons whipped cream; fold in half, then fold in half again. Arrange crepes on chilled serving plate and top with strawberry sauce. Serves 8.

Dessert Crepes

Flour 1½ cups
Sugar 1 tablespoon
Baking powder ½ teaspoon
Salt ½ teaspoon
Milk 2 cups
Eggs 2
Vanilla ½ teaspoon
Butter 2 tablespoons, melted
Butter 2 tablespoons

Sift together flour, sugar, baking powder, and salt. Pour milk, eggs, and vanilla into a blender and mix well. Slowly add dry ingredients and blend until smooth, then add 2 tablespoons of melted butter and mix until blended. Allow batter to stand 30 minutes before using. To make crepes, heat remaining 2 tablespoons of butter in shallow 6-inch cast-iron or aluminum skillet and swirl it around. When butter is completely melted and sizzling, wipe out pan with paper towel. Pour in about 2 tablespoons of batter and swirl this around until the bottom of the pan is covered. (Good crepes should be lacy looking, not thick like pancakes.) When batter is set, turn and cook other side. Continue until all the batter is used up. Makes 14 to 16 6-inch crepes.

Strawberry Supreme

Strawberries 3 pints, hulled and sliced
Sugar ¼ cup
Frozen orange juice concentrate 3 tablespoons, thawed
Rum 3 tablespoons
Whipping cream 1 cup, chilled
Sour cream 3 tablespoons
Sugar 2 tablespoons
Rum 2 teaspoons

Place berries in a large bowl. Combine sugar, orange juice concentrate, and rum, and pour over berries. Cover and chill for 1 to 2 hours. Whip cream lightly, then add sour cream, sugar, and rum and continue whipping until thick (do not overwhip). Spoon berries into 8 dessert dishes and top with whipped cream. Serves 8.

Strawberry-Orange Bavarian

Sugar   6 tablespoons
Unflavored gelatin   1¼-ounce envelope
Egg yolks   3
Orange liqueur   ¾ cup
Egg whites   3
Sugar   6 tablespoons
Whipping cream   1 cup, whipped
Strawberries   2 cups, hulled and sliced

Combine sugar with gelatin and mix well; set aside. Beat egg yolks slightly in the top of a double boiler and gradually add liqueur, stirring constantly; sprinkle with gelatin mixture and mix well. Cook over hot, but not boiling, water for 10 minutes or until mixture coats the back of a spoon; cool to room temperature. Meanwhile, beat egg whites, gradually adding sugar, until they are glossy. Fold in egg yolk mixture. Fold in whipped cream and pour into lightly oiled 5-cup mold. Chill until firm and unmold onto a serving plate; arrange berries over top and sides. Serves 8.

Strawberry-Rhubarb Crunch

Flour   1 cup, sifted
Oats   ¾ cup
Brown sugar   1 cup
Margarine   ½ cup, melted
Strawberries   2 cups
Rhubarb   2 cups chopped
Water   1 cup
Sugar   1 cup
Cornstarch   2 tablespoons
Vanilla   1 teaspoon

Mix together flour, oats, brown sugar, and margarine. Press ½ of mixture into a 9 by 9-inch pan; cover with berries and rhubarb. Combine water, sugar, and cornstarch in a saucepan and stir until blended; cook over medium heat until mixture thickens. Add vanilla and pour over strawberry-rhubarb mixture. Top with remaining crumb mixture. Bake at 350° for about 40 minutes. Serves 8.

Strawberry Shortcake

Flour 3 cups
Baking powder 3 teaspoons
Salt ½ teaspoon
Sugar 2 tablespoons
Shortening ½ cup
Egg 1, beaten
Milk ½ cup plus 2 tablespoons
Strawberries 2 cups, hulled and sliced
Whipping cream 1 cup, whipped

In a medium bowl, sift together flour, baking powder, salt, and sugar. Cut in shortening until mixture looks like small peas. Add egg and milk and stir gently. Spread mixture in a greased 8 by 8-inch pan and bake at 350° for 20 minutes, or until golden brown; cool. Cut into 6 sections; split each section in half. Divide 1 cup of berries over 6 halves, top with other halves, and divide up remaining berries. Top each shortcake with whipping cream. Serves 6.

Strawberry Torte

Strawberries 2½ cups, hulled
Sugar 1¼ cups
Water ⅓ cup
Light corn syrup 1 teaspoon
Egg whites 4
Whipping cream 1 pint, whipped
Ladyfingers 1 3-ounce package

Purée berries in a food processor or press through a strainer; set aside. Combine sugar, water, and corn syrup in a saucepan and boil over high heat until mixture reaches the soft ball stage, or 238°. Meanwhile, beat egg whites with an electric mixer until soft peaks are formed. Pour hot sugar mixture in a thin stream over egg whites, beating continually; continue to beat on high speed for 8 minutes more. Fold in strawberry purée, then fold in whipped cream. Line a 9-inch springform pan with split ladyfingers and pour in mixture. Wrap well with plastic wrap and freeze overnight. To serve, remove sides of pan and slide torte onto a serving plate. Serves 12.

Summer Cobbler

Sugar ½ cup
Flour 1 cup
Baking powder 2 teaspoons
Salt ¼ teaspoon
Milk ½ cup
Vanilla ½ teaspoon
Margarine 1 tablespoon, melted
Strawberries 1 cup, hulled and sliced
Blueberries 1 cup
Boiling water ¾ cup
Ice cream topping (optional)

In a large bowl, sift together sugar, flour, baking powder, and salt. Stir in milk, vanilla, and margarine. Spread batter in a greased 8 by 8-inch pan; sprinkle with berries. Carefully pour boiling water over top of cobbler. Bake at 375° for 25 to 30 minutes. Serve topped with ice cream if desired. Serves 9.

Summer Soufflé

Unflavored gelatin 2¼-ounce envelopes
Water ½ cup
Egg yolks 6
Sugar ¼ cup
Lemon yogurt 1 8-ounce container
Egg whites 6
Cream of tartar ½ teaspoon
Sugar ¼ cup
Whipping cream 1 cup, whipped until stiff
Strawberries 2 cups, hulled and sliced

Prepare a 1½ quart soufflé dish by extending a paper collar 2 inches above top of dish; secure with string around dish. In a small saucepan, soften gelatin in water; set aside. Combine egg yolks and sugar in the top of a double boiler and cook over medium high heat, stirring briskly, until thick. Melt gelatin over low heat and add to egg mixture. Stir in yogurt. Refrigerate about 45 minutes, or until mixture begins to thicken. In a large mixing bowl, beat egg whites with cream of tartar until they form soft peaks, then add sugar, a little at a time, and continue beating until mixture forms soft, shiny peaks. Whip cream until stiff. Fold whipped cream into egg yolk mixture and then gently fold in egg whites. Fold in berries. Pour into prepared dish and chill for at least 4 hours. Remove paper collar and serve. Serves 6 to 8.

Raspberry Truffles

Semisweet chocolate chips   1 6-ounce package
Butter   6 tablespoons, cut in chunks
Lingonberry preserves   ½ cup
Brandy   2 tablespoons
Fresh raspberries   16
Unsweetened cocoa   ½ cup, sifted

Melt chocolate and butter in the top of a double boiler over hot, but not boiling, water. Pour into a small bowl and cool to room temperature. Push lingonberry preserves through a wire strainer and discard pulp. Add preserves and brandy to chocolate mixture and stir well; refrigerate overnight. Scrape pieces from hardened chocolate mixture with a teaspoon and wrap around each berry. Refrigerate berries on a wax paper–lined cookie sheet for 1 hour; roll each berry in cocoa. Makes 16 truffles.

Chocolate-Dipped Strawberries

Semisweet chocolate chips   1 6-ounce package
Perfect fresh strawberries   20, with leaves attached

Line a cookie sheet with waxed paper. Slowly melt chocolate in a double boiler over hot, but not boiling, water. Dip each berry into melted chocolate by holding onto stem and leaves; place on prepared cookie sheet. Chill in refrigerator until firm. Use as a garnish or as a light dessert. Makes 20 confections.

U.S. and Metric Measurements

Approximate conversion formulas are given below for commonly used U.S. and metric kitchen measurements.

Teaspoons	x	5	= milliliters
Tablespoons	x	15	= milliliters
Fluid ounces	x	30	= milliliters
Fluid ounces	x	0.03	= liters
Cups	x	240	= milliliters
Cups	x	0.24	= liters
Pints	x	0.47	= liters
Dry pints	x	0.55	= liters
Quarts	x	0.95	= liters
Dry quarts	x	1.1	= liters
Gallons	x	3.8	= liters
Ounces	x	28	= grams
Ounces	x	0.028	= kilograms
Pounds	x	454	= grams
Pounds	x	0.45	= kilograms
Milliliters	x	0.2	= teaspoons
Milliliters	x	0.07	= tablespoons
Milliliters	x	0.034	= fluid ounces
Milliliters	x	0.004	= cups
Liters	x	34	= fluid ounces
Liters	x	4.2	= cups
Liters	x	2.1	= pints
Liters	x	1.82	= dry pints
Liters	x	1.06	= quarts
Liters	x	0.91	= dry quarts
Liters	x	0.26	= gallons
Grams	x	0.035	= ounces
Grams	x	0.002	= pounds
Kilograms	x	35	= ounces
Kilograms	x	2.2	= pounds

Temperature Equivalents

Fahrenheit	− 32	x 5	÷ 9	= Celsius
Celsius	x 9	÷ 5	+ 32	= Fahrenheit

U.S. Equivalents

1 teaspoon	= ⅓ tablespoon
1 tablespoon	= 3 teaspoons
2 tablespoons	= 1 fluid ounce
4 tablespoons	= ¼ cup or 2 ounces
5⅓ tablespoons	= ⅓ cup or 2⅔ ounces
8 tablespoons	= ½ cup or 4 ounces
16 tablespoons	= 1 cup or 8 ounces
⅜ cup	= ¼ cup plus 2 tablespoons
⅝ cup	= ½ cup plus 2 tablespoons
⅞ cup	= ¾ cup plus 2 tablespoons
1 cup	= ½ pint or 8 fluid ounces
2 cups	= 1 pint or 16 fluid ounces
1 liquid quart	= 2 pints or 4 cups
1 liquid gallon	= 4 quarts

Metric Equivalents

1 milliliter	= 0.001 liter
1 liter	= 1000 milliliters
1 milligram	= 0.001 gram
1 gram	= 1000 milligrams
1 kilogram	= 1000 grams

Index

Other Cookbooks from Pacific Search Press

The Apple Cookbook by Kyle D. Fulwiler
Asparagus: The Sparrowgrass Cookbook by Autumn Stanley
Bone Appétit! Natural Foods for Pets by Frances Sheridan Goulart
The Carrot Cookbook by Ann Saling
The Crawfish Cookbook by Norma S. Upson
The Dogfish Cookbook by Russ Mohney
The Eggplant Cookbook by Norma S. Upson
The Green Tomato Cookbook by Paula Simmons
*Mushrooms 'n Bean Sprouts: A First Step for Would-be
    Vegetarians* by Norma M. MacRae, R.D.
My Secret Cookbook by Paula Simmons
The Natural Fast Food Cookbook by Gail L. Worstman
Rhubarb Renaissance: A Cookbook by Ann Saling
The Salmon Cookbook by Jerry Dennon
Starchild & Holahan's Seafood Cookbook by Adam Starchild
    and James Holahan
The Whole Grain Bake Book by Gail L. Worstman
Wild Mushroom Recipes by Puget Sound Mycological Society
The Zucchini Cookbook by Paula Simmons